Discrete Transforms

Discrete Transforms

Jean M. Firth

Senior Lecturer in Engineering Mathematics,
University of Newcastle upon Tyne, Newcastle, UK

Springer-Science+Business Media, B.V.

First edition 1992

© 1992 Jean M. Firth
Originally published by Chapman & Hall in 1992

Typeset in 10/12pt Times by
Thomson Press (India) Ltd, New Delhi, India

ISBN 978-0-412-42990-3   ISBN 978-94-011-2358-7 (eBook)
DOI 10.1007/978-94-011-2358-7

A catalogue record for this book is available from the British Library

Library of Congress Cataloging-in-Publication data

Firth, Jean M., 1938–
    Discrete transforms / Jean M. Firth.
        p.   cm.
    Includes bibliographical references and index.

    1. Transformations (Mathematics)   I. Title.
QA601.F53   1992
515'.723—dc20
                                        91-44211
                                        CIP

Contents

Acknowledgements

The author wishes to thank Dr Christopher J.S. Petrie for his contribution to Figures 2.6, 3.13, 3.14, 3.18, 3.20, 3.22, 3.23 and 3.24.

Preface

The analysis of signals and systems using transform methods is a very important aspect of the examination of processes and problems in an increasingly wide range of applications. Whereas the initial impetus in the development of methods appropriate for handling discrete sets of data occurred mainly in an electrical engineering context (for example in the design of digital filters), the same techniques are in use in such disciplines as cardiology, optics, speech analysis and management, as well as in other branches of science and engineering.

This text is aimed at a readership whose mathematical background includes some acquaintance with complex numbers, linear differential equations, matrix algebra, and series. Specifically, a familiarity with Fourier series (in trigonometric and exponential forms) is assumed, and an exposure to the concept of a continuous integral transform is desirable. Such a background can be expected, for example, on completion of the first year of a science or engineering degree course in which transform techniques will have a significant application. In other disciplines the readership will be past the second year undergraduate stage. In either case, the text is also intended for earlier graduates whose degree courses did not include this type of material and who now find themselves, in a professional capacity, requiring a knowledge of discrete transform methods.

Rapid changes in undergraduate courses have meant that material concerning the analysis of signals, continuous and discrete, now appears at an earlier stage than once might have been the case, or is included for the first time at undergraduate level on a compulsory basis. How the material is introduced varies between institutions: topics might be taught by departments of computing and mathematics, or 'in house' by the 'user'departments. Pressures on timetables determine the approach to a certain extent, but it is desirable that the syllabus provide an understanding of a transform and its properties before, or at least in parallel with, consideration of any particular application.

The literature is extensive and excellent references are available. However, the balance of a book is often tipped in favour of a particular field of application (in which case the treatment of transform methods as such is sometimes too condensed or too advanced to be accessible to the student, who might in any case be interested in quite other applications). Our present purpose is to provide an introduction to the more advanced or specialized literature in which the emphasis is on the development of transform techniques. Although applications are illustrated, the primary intention is to describe how transforms and their inverses are derived, how different transforms are related, how transforms can be used to handle discrete data sets, and then to consider what efficient computational methods are available to carry out the associated digital calculation processes.

The text is based on a course provided by the Department of Engineering Mathematics at the University of Newcastle upon Tyne at the request of the Department of Electrical and Electronic Engineering. Until the early 1980s, that department had included such topics as Z-transforms and fast Fourier transforms in fairly condensed form within those of its 'own' courses in which these techniques are applied, but concluded that a greater amount of time should be spent on this material, at an earlier stage in the degree course. In its present form the course is taken by all second year students reading degrees in electrical and electronic engineering and in microelectronics and software engineering. It is also available to undergraduates in other years, in other disciplines, and to postgraduates.

In Chapter 1, we rehearse the application of the Fourier series representation of a function defined on a finite range to determination of the amplitude and phase spectra of such a signal. In the remainder of that chapter (and in Appendix A) we proceed to consider the application of integral transforms to (continuous-time) signals defined on an infinite or semi-infinite range. In Chapter 2, the Fourier transform is considered in further detail. These two chapters are intended to consolidate and extend any prior knowledge of continuous-time systems, with particular reference to the concept of 'convolution'.

Chapters 3 and 4 relate to the discrete version of the Laplace transform, known as the Z-transform, and its applications. Chapter 5 is

concerned with the discrete version of the Fourier transform. In
Chapter 6 we consider the description of the discrete Fourier transform
in terms of matrices, and anticipate the 'fast' computation algorithms
established in (the concluding) Chapter 7.

The problem sets at the end of each chapter do not consist merely
of repetitious exercises but include applications and also develop-
mental material which anticipates later work. Although the latter is
not compromised if those problems are omitted, understanding is
assisted if they are not. With very few exceptions, solutions are given
within the statement of the problem, and guidance is included where
this seems useful.

J. M. Firth
Department of Engineering Mathematics
Faculty of Engineering
University of Newcastle upon Tyne

1

Fourier series, integral theorem, and transforms: a review

A continuous-time system may be described by differential, difference or algebraic equations. In other applications, the independent variable might be a spatial dimension, rather than time. It is often the case that when examining either signals or the behaviour of a system, the use of harmonic analysis and transform methods provides valuable information. An obvious example is provided if we consider vibrations – which might be positively unwelcome. In a mechanical system, vibrations can be potentially or actually destructive. The component parts have their own natural frequencies of vibration, and internal or external 'noise' might produce oscillations of an unacceptably large amplitude. This is called 'resonance', and if the source is identified then corrective measures can be considered, such as redesigning the damping characteristics of an affected component or (if possible) removing the cause of the noise. Similar factors are important in the design of the speakers and amplifiers of an electrical system.

As indicated in the preface to this book, it is assumed that the reader has some familiarity with the Fourier series representation of a time-domain signal, say $x(t)$, with the concepts of integral transforms as evinced by the Laplace transform of a signal, say $F(s) = \mathscr{L}\{x(t)\}$, and has some appreciation of the significance of convolution.

The purpose of this chapter is to review that material, and to highlight those aspects which are most relevant to the development of the subject. At the same time, there will be introduced, gradually, material which might be new to some readers – in particular the concept of a 'system transfer function'.

1.1 FOURIER SERIES

A Fourier series is the representation of a function $x(t)$ by an infinite series of terms of orthogonal functions. If these are the trigonometric functions then the representation of a function $x(t)$ defined on an interval $[a, b]$ takes the form

$$x(t) \sim \tfrac{1}{2}a_0 + \sum_{n=1}^{\infty} \left\{ a_n \cos\left(\frac{2n\pi t}{b-a}\right) + b_n \sin\left(\frac{2n\pi t}{b-a}\right) \right\} \qquad (1.1)$$

in which

$$a_n + jb_n = \frac{2}{b-a} \int_a^b x(t') e^{j2n\pi t'/(b-a)} \, dt' \qquad (1.2)$$

(Equation (1.2) combines the 'usual' equations

$$a_n = \frac{2}{b-a} \int_a^b x(t') \cos\left(\frac{2n\pi t'}{b-a}\right) dt'$$

and

$$b_n = \frac{2}{b-a} \int_a^b x(t') \sin\left(\frac{2n\pi t'}{b-a}\right) dt'$$

and indicates that there might be computational efficiency to be gained by using the identity $e^{j\theta} = \cos\theta + j\sin\theta$.)

At this point we emphasize that, whether or not $x(t)$ is defined outside the interval $[a, b]$, and whether or not $x(t)$ is a periodic function, *the Fourier series representation is periodic*, because of its trigonometric nature. Its sum at any point t is $\tfrac{1}{2}\{x(t^-) + x(t^+)\}$ which reduces to $x(t)$ without ambiguity at points where $x(t)$ is continuous.

Equation (1.1) can be rewritten as

$$x(t) \sim \tfrac{1}{2}c_0 + \sum_{n=1}^{\infty} c_n \cos\left(\frac{2n\pi t}{b-a} - \theta_n\right) \qquad (1.3)$$

in which

$$c_n = \sqrt{a_n^2 + b_n^2}$$

and

$$\theta_n = \tan^{-1}\left(\frac{b_n}{a_n}\right)$$

This is very informative, in practice. The discrete integer variable n denotes a particular frequency and so a plot of amplitude $\{c_n\}$ against n is an indicator of what frequencies are present in a signal and, for example, which are the *dominant* components. By this we mean frequencies associated with relatively large amplitudes, of which there might be several. Equally, the phase spectrum of $\{\theta_n\}$ against n indicates to what extent the harmonic components are or are not *in phase*. Components in phase have common zeros, otherwise we may speak of phase differences using such terms as 'shift', 'lag' or 'delay' (in time). (When computing $\theta_n = \tan^{-1}(b_n/a_n)$, care should be taken to assign θ_n to the correct quadrant in some principal range, be that $0 \leqslant \theta < 2\pi$ or $-\pi < \theta \leqslant \pi$, for example.)

1.2 FOURIER EXPONENTIAL SERIES

If we take the interval $[a, b]$ to be $[-l, l]$ and convert from trigonometric functions to the exponential equivalent, then the previous expressions can be replaced by the series

$$x(t) \sim \sum_{-\infty}^{\infty} X_n e^{jn\pi t/l} \tag{1.4}$$

in which

$$X_n = \frac{1}{2l} \int_{-l}^{l} x(t') e^{-jn\pi t'/l} \, dt' \tag{1.5}$$

(Because of the periodic nature of the series we could integrate over *any* interval $[a, b] = [a, a + 2l]$.) Information about amplitude and phase is contained in the coefficient X_n, in general a complex number.

$|X_n|$ provides the amplitude spectrum and $\arg X_n$ provides the phase spectrum.

Worked Example 1.1

The rectified sine wave

$$x(t) = \frac{\pi}{2} \left| \sin\left(\frac{2\pi t}{T}\right) \right| \qquad 0 < t < T$$

is sketched in Fig. 1.1. Find its Fourier exponential series.

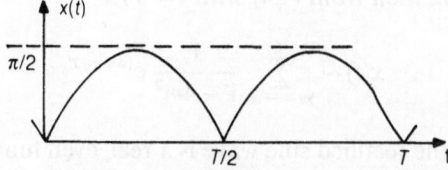

Fig. 1.1 The rectified sine wave.

Solution

Integrating over $[a, a + 2l] = [0, T]$, Equation (1.5), appropriately modified, gives

$$X_n = \frac{\pi}{2}\frac{1}{T}\left\{\int_0^{T/2} \sin\left(\frac{2\pi t'}{T}\right)e^{-j2n\pi t'/T}\,dt' \right.$$
$$\left. + \int_{T/2}^{T}\left[-\sin\left(\frac{2\pi t'}{T}\right)\right]e^{-j2n\pi t'/T}\,dt'\right\}$$

If in the second integral we put $\tau = t' - T/2$ then, noting that $\sin(\theta + \pi) = -\sin\theta$ and $e^{-jn\pi} = (-1)^n$, we obtain

$$\int_0^{T/2}(-1)^n \sin\left(\frac{2\pi\tau}{T}\right)e^{-j2n\pi\tau/T}\,d\tau$$

which, apart from the factor $(-1)^n$, is the first integral. Then

$$X_n = \frac{\pi}{2T}[1 + (-1)^n]\int_0^{T/2}\sin\left(\frac{2\pi t'}{T}\right)e^{-j2n\pi t'/T}\,dt'$$

It follows that if n is an *odd* integer, $X_n = 0$. For *even* values of n, we may integrate using trigonometric relationships or use the identity $\sin\theta = (e^{j\theta} - e^{-j\theta})/2j$. Using the latter,

$$X_n = \frac{\pi}{T}\int_0^{T/2}\frac{1}{2j}\{e^{j2\pi t'(1-n)/T} - e^{-j2\pi t'(1+n)/T}\}\,dt'$$
$$= \frac{\pi}{2Tj}\frac{T}{2\pi j}\left[\frac{e^{j2\pi t'(1-n)/T}}{1-n} + \frac{e^{-j2\pi t'(1+n)/T}}{1+n}\right]_0^{T/2}$$

Since $e^{j\pi(1-n)} = e^{-j\pi(1+n)} = -1$, when n is even, this expression simplifies to

$$X_n = \frac{1}{1-n^2}$$

Putting $n = 2m$, then from (1.4) with $l = T/2$

$$x(t) \sim \sum_{m=-\infty}^{\infty}\frac{1}{1-4m^2}e^{j4m\pi t/T}$$

(Observe that the rectified sine wave is a real, even function of t. The coefficient X_n obtained is a real, even function of n.)   ●

Worked Example 1.2

Given $x(t) = e^t$, $-\pi < t < \pi$, obtain the Fourier exponential series for $x(t)$ and illustrate the frequency spectrum.

Solution

From (1.5) we have

$$X_n = \frac{1}{2\pi} \int_{-\pi}^{\pi} e^{t'(1-jn)} \, dt' = \frac{1}{2\pi} \left[\frac{e^{t'(1-jn)}}{1-jn} \right]_{-\pi}^{\pi}$$

Noting that $e^{\pm jn\pi} = (-1)^n$ and that $e^{\pi} - e^{-\pi} = 2\sinh\pi$,

$$X_n = \frac{\sinh\pi}{\pi} \frac{(-1)^n}{1-jn}$$

and so (1.4) gives

$$x(t) \sim \frac{\sinh\pi}{\pi} \sum_{-\infty}^{\infty} \frac{(-1)^n}{1-jn} e^{jnt}$$

The modulus of X_n is

$$|X_n| = \frac{\sinh\pi}{\pi\sqrt{1+n^2}}$$

which is an even function of n with a maximum at $n = 0$, hence the amplitude spectrum is as shown in Fig. 1.2.

In terms of its real and imaginary parts

$$X_n = (-1)^n \frac{\sinh\pi}{\pi(1+n^2)} + j(-1)^n \frac{n\sinh\pi}{\pi(1+n^2)}$$

First suppose $n > 0$. It is seen that the real and imaginary parts of X_n are either both positive or both negative. The angle $\theta_n = \tan^{-1} n$

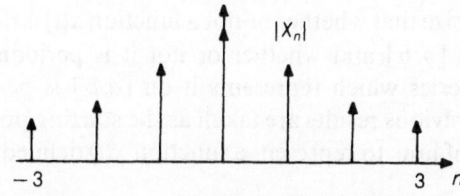

Fig. 1.2 The amplitude spectrum of $x(t)$.

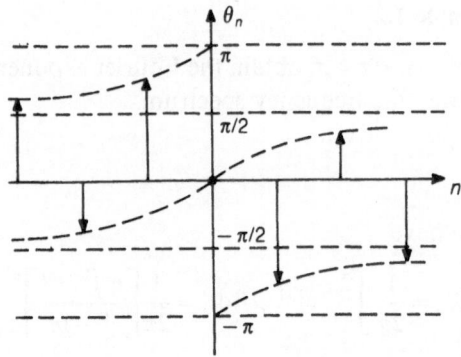

Fig. 1.3 The phase spectrum of $x(t)$.

is therefore assigned to the first quadrant if n is even and to the third quadrant if n is odd. If the principal value range is taken to be $(-\pi, \pi)$ and if $\alpha_n = \tan^{-1} n$ is an acute angle, then we have

$$\theta_n = \begin{cases} \alpha_n & n \text{ even} \\ \alpha_n - \pi & n \text{ odd} \end{cases}$$

in which α_n is the acute (first quadrant) angle.

Now suppose $n < 0$. The real and imaginary parts of X_n are of opposite sign. If n is even, θ_n is assigned to the fourth quadrant and we may write $\theta_n = -\alpha_n$. Similarly, if n is odd and negative, $\theta_n = \pi - \alpha_n$. Hence the phase spectrum given in Fig. 1.3.   ●

Equations (1.4) and (1.5) can be regarded as constituting a *transform pair*. The quantity X_n (a function of the *discrete* variable n, which takes only integer values), can be regarded as an *integral transform* of $x(t)$. Equation (1.4) represents an *inversion* process, inasmuch as if we have an expression for X_n, then we can recover $x(t)$.

1.3 THE FOURIER INTEGRAL THEOREM

We re-emphasize that whether or not a function $x(t)$ is defined outside a finite range $[a, b]$ and whether or not it is periodic, the sum of the Fourier series which represents it on $[a, b]$ *is* periodic. In this section, the previous results are taken as the starting point to address the question of how to represent a function $x(t)$ defined on an *infinite* range.

Substituting (1.5) into (1.4) gives

$$x(t) = \frac{1}{2l} \sum_{n=-\infty}^{\infty} \left\{ \int_{-l}^{l} x(t')e^{-jn\pi t'/l}\,dt' \right\} e^{jn\pi t/l} \tag{1.6}$$

In this, the function $x(t)$ is defined on the finite range $[-l, l]$. The parameter n represents discrete frequencies. Consider increasing the range, letting $l \to \infty$. As $[-l, l] \to (-\infty, \infty)$, a limit process takes us to the equivalent of (1.6) for a function $x(t)$ defined on a doubly infinite range, and this takes the form

$$x(t) = \frac{1}{2\pi} \int_{-\infty}^{\infty} \left\{ \int_{-\infty}^{\infty} x(t')e^{-jwt'}\,dt' \right\} e^{jwt}\,dw \tag{1.7}$$

(Should this not have been encountered before, an outline proof is given in Appendix A.)

In (1.7) the variable w describes frequency, but that is now regarded as a *continuous* variable. If frequency is denoted by f and measured in hertz, substituting $w = 2\pi f$ (so that $dw = 2\pi df$) in (1.7) leads to the more symmetrical form

$$x(t) = \int_{-\infty}^{\infty} \left\{ \int_{-\infty}^{\infty} x(t')e^{-j2\pi ft'}\,dt' \right\} e^{j2\pi ft}\,df \tag{1.8}$$

Either (1.7) or (1.8) is a statement of the Fourier integral theorem, which is the basis of integral transform theory.

An alternative form of (1.7) is given if we substitute $-w$ for w on the half-range $-\infty < w < 0$. We obtain

$$x(t) = \frac{1}{2\pi} \int_{-\infty}^{0} \left\{ \int_{-\infty}^{\infty} x(t')e^{+jwt'}\,dt' \right\} e^{-jwt}(-dw)$$
$$+ \frac{1}{2\pi} \int_{0}^{\infty} \left\{ \int_{-\infty}^{\infty} x(t')e^{-jwt'}\,dt' \right\} e^{-jwt}\,dw$$

Using the minus sign to exchange the limits of integration in the first term and noting that $\cos\theta = (e^{j\theta} + e^{-j\theta})/2$, this reduces to

$$x(t) = \frac{1}{\pi} \int_{0}^{\infty} \left\{ \int_{-\infty}^{\infty} x(t')\cos[w(t - t')]\,dt' \right\} dw \tag{1.9}$$

Similarly, the second integration in (1.8) can be rewritten as an integral over $0 \leqslant f < \infty$.

1.4 ODD AND EVEN FUNCTIONS

If $x(t)$ is an even function so that $x(-t') = x(t')$, then (1.9) can be further simplified by substituting $-t'$ for t' on the half-range $-\infty < t' < 0$. This gives

$$x(t) = \frac{1}{\pi} \int_0^\infty \left\{ \int_{-\infty}^0 x(-t') \cos\left[w(t+t')\right](-dt') \right.$$
$$\left. + \int_0^\infty x(t') \cos\left[w(t-t')\right] dt' \right\} dw$$

Expanding the cosine terms and noting that $x(t)$ is even, we obtain the result

$$x(t) = \frac{2}{\pi} \int_0^\infty \left\{ \int_0^\infty x(t') \cos(wt') dt' \right\} \cos(wt) dw \qquad (1.10)$$

In the same way, if $x(t)$ is odd then $x(-t') = -x(t')$ and the corresponding result for odd functions takes the form

$$x(t) = \frac{2}{\pi} \int_0^\infty \left\{ \int_0^\infty x(t') \sin(wt') dt' \right\} \sin(wt) dw \qquad (1.11)$$

The implication here is that if a function is defined on the half-range $0 \leqslant t < \infty$, or if our interest is confined to that range only, we may choose to treat it as either odd or even (whether or not it is). This is analogous to the idea of representing a function by a half-range Fourier *series* (of cosine terms only or of sine terms only) on the finite half-range $[0, l]$.

The integrals involved in previous expressions are improper integrals, having infinite limits, and so do not necessarily exist. It is possible that a 'one-sided' integral over the range $[0, \infty)$ converges when the 'two-sided' integral over $(-\infty, \infty)$ does not, but the question of existence must be born in mind in any application.

1.5 THE FOURIER TRANSFORM

If in the double-integral versions of the Fourier integral theorem, (1.7)–(1.11), we take the 'inner' integral as the definition of some integral transform, then the double integral can be replaced by a transform pair.

In particular, if we define the function

$$X(f) = \int_{-\infty}^{\infty} x(t')e^{-j2\pi f t'}\,dt' = \mathscr{F}\{x(t)\} \tag{1.12}$$

to be the Fourier transform of the time-domain signal $x(t)$, then substitution into (1.8) gives the inversion integral

$$x(t) = \int_{-\infty}^{\infty} X(f)e^{j2\pi f t}\,df \tag{1.13}$$

whereby $x(t)$ can be recovered from a known transform $X(f)$. As given in (1.12), $X(f)$ is referred to as the *two-sided* Fourier transform of $x(t)$, and we have already indicated that $x(t)$ has no such transform if this integral does not converge. With $x(t) = 0$ on $t < 0$ we may define the *one-sided* transform to be

$$X(f) = \int_{0}^{\infty} x(t')e^{-j2\pi f t'}\,dt' \tag{1.14}$$

This is inverted by (1.13) without any change to the limits of integration, which remains an integral over $-\infty < f < \infty$.

In all three cases, convergence requires that the integral of the modulus of the integrand exist. In some cases establishing an expression for $X(f)$ by direct use of the defining integral presents problems, even if some form of limiting process is invoked, but, as will be seen in Chapter 2, it is often possible to make use of properties of the Fourier transform itself to resolve such difficulties.

Worked Example 1.3

Find the Fourier transform of the function defined as follows:

$$x(t) = \begin{cases} 0 & |t| > 1 \\ 1 + t & -1 < t < 0 \\ 1 - t & 0 < t < 1 \end{cases}$$

Solution

From (1.12)

$$X(f) = \int_{-1}^{0} (1 + t')e^{-j2\pi f t'}\,dt' + \int_{0}^{1} (1 - t')e^{-j2\pi f t'}\,dt'$$

If t' is replaced by $-t'$ in the first integral,

$$X(f) = 2 \int_0^1 (1 - t') \cos(2\pi f t') \, dt'$$

$$= 2 \left[(1 - t') \frac{\sin(2\pi f t')}{2\pi f} - \frac{\cos(2\pi f t')}{(2\pi f)^2} \right]_0^1$$

$$= \frac{1}{2\pi^2 f^2} [1 - \cos(2\pi f)]$$

From the identity

$$\cos(2\theta) = 1 - 2\sin^2\theta$$

it follows that

$$X(f) = \left[\frac{\sin(\pi f)}{\pi f} \right]^2$$

(Note that $x(t)$ and $X(f)$ are real, even functions of t and f, respectively.) ●

Worked Example 1.4

Find the one-sided Fourier transform of the function

$$x(t) = e^{-t} \cos t \qquad t \geqslant 0$$

Solution

Writing the cosine factor in $x(t)$ in exponential form, (1.14) gives

$$X(f) = \int_0^\infty \tfrac{1}{2} \{ e^{-(1-j+j2\pi f)t'} + e^{-(1+j+j2\pi f)t'} \} \, dt'$$

$$= \frac{1}{2} \left[\frac{1}{1 + j(2\pi f - 1)} + \frac{1}{1 + j(2\pi f + 1)} \right]$$

after integration and substituting limits. With appropriate conjugation to obtain a real denominator, this simplifies to

$$X(w) = \frac{(2 + w^2) - jw^3}{4 + w^4}$$

in which we have put $w = 2\pi f$ to obtain a less cumbersome expression. ●

1.6 THE FOURIER SINE AND COSINE TRANSFORMS

All Fourier transform pairs can be written in terms of either f or w. For example, we may define the (one-sided) Fourier cosine transform to be

$$X_c(w) = \int_0^\infty x(t')\cos(wt')\,dt' = \mathscr{F}_c\{x(t)\} \tag{1.15}$$

Then, from (1.10), the inverse is

$$x(t) = \frac{2}{\pi}\int_0^\infty X_c(w)\cos(wt)\,dw \tag{1.16}$$

(It should be noted that in some texts any constant multiplier in the integral theorem (in this case $2/\pi$) might be included in the definition of the transform rather than in the inversion integral. Alternatively, in order that the transform pair appear to be symmetrical, a factor $\sqrt{2/\pi}$ might appear as a multiplier in each.)

Similarly, we may define the (likewise one-sided) Fourier sine transform to be

$$X_s(w) = \int_0^\infty x(t')\sin(wt')\,dt' = \mathscr{F}_s\{x(t)\} \tag{1.17}$$

Then it follows from (1.11) that

$$x(t) = \frac{2}{\pi}\int_0^\infty X_s(w)\sin(wt)\,dw \tag{1.18}$$

These results, at this stage, are given for illustrative purposes only (as the main objective of this book is to consider the Fourier and Laplace transforms and their discrete forms), and are included to amplify the earlier remark concerning the central position of the integral theorem in transform theory. However, Problems 2.10 and 2.11 demonstrate some of the properties of the cosine and sine transforms.

1.7 THE LAPLACE TRANSFORM

The definition of the Laplace transform as the one-sided integral

$$F(s) = \int_0^\infty x(t')e^{-st'}\,dt' = \mathscr{L}\{x(t)\} \tag{1.19}$$

will be familiar, but perhaps less familiar is its inverse

$$x(t) = \frac{1}{2\pi j} \int_{c-j\infty}^{c+j\infty} F(s) e^{st} \, ds \qquad (1.20)$$

The latter is a complex integral in the complex plane of s. The constant c is real and positive. To show that (1.19) and (1.20) do indeed form a transform pair, we can combine them and compare the result with the integral theorem. So as not to prejudge the issue, we will initially refer to the right-hand side of (1.20) as the function $I(t)$. By putting $s = c + jy$ and at the same time substituting for $F(s)$ as defined in (1.19), we obtain

$$I(t) = \frac{1}{2\pi j} \int_{-\infty}^{\infty} \left\{ \int_{0}^{\infty} x(t') e^{-(c+jy)t'} \, dt' \right\} e^{(c+jy)t} (j \, dy)$$

$$= \frac{e^{ct}}{2\pi} \int_{-\infty}^{\infty} \left\{ \int_{0}^{\infty} [x(t') e^{-ct'}] e^{-jyt'} \, dt' \right\} e^{jyt} \, dy$$

Comparing this with the integral theorem in the form

$$x(t) = \frac{1}{2\pi} \int_{-\infty}^{\infty} \left\{ \int_{-\infty}^{\infty} x(t') e^{-jwt'} \, dt' \right\} e^{jwt} \, dw$$

where the latter refers to a function $x(t)$ defined on $-\infty < t < \infty$, the expression for $I(t)$ involves instead a function $e^{ct}[x(t)e^{-ct}]$ for $t \geqslant 0$, and zero for $t < 0$.

It follows that

$$I(t) = \begin{cases} x(t) & \text{if } t \geqslant 0 \\ 0 & \text{if } t < 0 \end{cases}$$

and so the complex integral in (1.20) does indeed invert the Laplace transform.

There is not, in the Laplace transform pair, the symmetry seen in the Fourier transform and its inverse ((1.12) and (1.13)). The symmetry of the latter leads to a powerful property of the Fourier transform which is not shared by the Laplace transform, as will be seen in Chapter 2, and which allows us to establish expressions for the Fourier transform not readily available from its integral definition. As far as the *one*-sided Fourier transform is concerned, if it exists then it can be obtained from the Laplace transform. For example, in terms of w, we can deduce that

$$X(w) = \int_0^\infty x(t')e^{-jwt'}\,dt' = F(jw)$$

on comparison with (1.19).

However, the existence of the Laplace transform does *not* imply that the function also has a Fourier transform.

1.8 LAPLACE TRANSFORM PROPERTIES AND PAIRS

As it is assumed that readers are reasonably familiar with the Laplace transform, it is not proposed to rehearse its properties in any great detail.

On the question of *existence*, however, in Section 1.7 it was indicated that the Laplace parameter s should be regarded as a complex quantity, $s = c + jy$, in which c is real and positive. It follows that, in

Table 1.1 Laplace transform properties, $F(s) = \mathscr{L}\{x(t)\}$

Property	Time domain	Transform domain
L1 First shift theorem	$e^{-at}x(t)$	$F(s+a)$
L2 Second shift theorem*	$x(t-a)H(t-a)$	$e^{-as}F(s), a \geqslant 0$
L3 Differentiation in time	(i) $\dfrac{dx}{dt}$	$sF(s) - x(0)$
	(ii) $\dfrac{d^n x}{dt^n}$	$s^n F(s) - \displaystyle\sum_{k=1}^{n} s^{k-1}x^{(n-k)}(0)$
L4 Multiplication by t	$tx(t)$	$-\dfrac{dF}{ds}$
L5 Periodic functions	$x(t) = x(t+nT)$	$\dfrac{1}{1-e^{-sT}}\displaystyle\int_0^T x(t')e^{-st'}\,dt'$
L6 Convolution	$\displaystyle\int_0^t x_1(t')x_2(t-t')\,dt'$	$F_1(s)F_2(s)$
	$= \displaystyle\int_0^t x_1(t-t')x_2(t')\,dt'$	

*The notation $H(t-a)$ is used for the unit step function defined by

$$H(t-a) = \begin{cases} 0 & t < a \\ 1 & t \geqslant a \end{cases}$$

Alternative notations include, for example, $u_a(t)$.

the definition of the Laplace transform, the integrand includes an exponential 'damping factor', so that

$$\mathscr{L}\{x(t)\} = \int_0^\infty x(t') e^{-ct'} e^{-jyt'}\, dt'$$

This is why a function which is *exponentially bounded* (meaning that $x(t)e^{-ct} \to 0$ as $t \to \infty$, or $|x(t)| < Me^{kt}$ where M is a constant and $c > k$) has a Laplace transform, whereas the Fourier transform might not exist.

Table 1.1 is a short selection of some of the properties of the Laplace transform. Worked examples are included showing proofs, but in general it is assumed that the reader can verify results either by using the integral definition (1.19) or by using properties already proved.

For present purposes, the following worked examples are offered merely as a reminder as to how one might use the defining integral of the Laplace transform, or properties already established.

Worked Example 1.5

Prove property L2 in Table 1.1, that is, that

$$\mathscr{L}\{x(t-a)H(t-a)\} = e^{-as}F(s)$$

Solution

From the defining integral, (1.19), the right-hand side of the above expression can be written as follows:

$$\int_0^\infty x(t') e^{-s(a+t')}\, dt'$$

which, on substituting $a + t' = t$, becomes

$$\int_a^\infty x(t-a)e^{-st}\, dt$$

Since the unit function $H(t-a) = 0$ if $t < a$, we may introduce it as a factor in the integrand and replace the lower limit by zero to obtain

$$\int_0^\infty \{x(t-a)H(t-a)\}e^{-st}\, dt$$

which, by definition, is the Laplace transform of

$$x(t-a)H(t-a) \qquad\qquad\qquad\bullet$$

Worked Example 1.6

If $\mathcal{L}\{x(t)\} = F(s)$, obtain the Laplace transform of the modulated signal $\sin(wt)x(t)$.

Solution

The Laplace Transform is *linear*, in that if the transforms of $x_1(t)$ and $x_2(t)$ exist, then from the integral definition it follows that

$$\mathcal{L}\{a_1 x_1(t) + a_2 x_2(t)\} = a_1 F_1(s) + a_2 F_2(s)$$

We can then use the first shift theorem (property L1 in Table 1.1) to write

$$\mathcal{L}\{\sin(wt)x(t)\} = \frac{1}{2j}[\mathcal{L}\{e^{jwt}x(t)\} - \mathcal{L}\{e^{-jwt}x(t)\}]$$

$$= \frac{1}{2j}[F(s - jw) - F(s + jw)] \qquad \bullet$$

Table 1.2 Laplace transform pairs

	$x(t)$	$F(s)$
LT1	1	$1/s$
LT2	$H(t - a)$	e^{-as}/s
LT3	$t^n/n!$	$1/s^{n+1}$
LT4	e^{at}	$1/(s - a)$
LT5	$\cos(at)$	$s/(s^2 + a^2)$
LT6	$\sin(at)$	$a/(s^2 + a^2)$
LT7	$\cosh(at)$	$s/(s^2 - a^2)$
LT8	$\sinh(at)$	$a/(s^2 - a^2)$
LT9*	$\delta(t - a)$	e^{-as}

*The notation $\delta(t - a)$ is used to describe a unit impulse occurring at the instant $t = a$. The impulse function can be defined in various ways, for example

$$\delta(t - a) = \lim_{\varepsilon \to 0}[H(t - a) - H(t - a - \varepsilon)]\frac{1}{\varepsilon}$$

Of its various properties, one of the most important is the *sifting* property, which can be stated as follows:

$$\int_{-\infty}^{t} x(t')\delta(t' - a)\,dt' = H(t - a)x(a).$$

Laplace transform properties have their equivalents when we consider the properties of other transforms. (In particular, those known as *shift*, *delay* or *translation* properties.) From the integral definition and properties can be obtained a table of Laplace transform pairs, a selection of which is given in Table 1.2.

The next two examples illustrate applications of the results of Tables 1.1 and 1.2.

Worked Example 1.7

Use the differentiation property L3 of Table 1.1 to show that if $F(s) = \mathcal{L}\{x(t)\}$ then

$$\mathcal{L}\left\{\int_0^t x(t')\,dt'\right\} = \frac{F(s)}{s}$$

Use this to solve the circuit equation

$$L\frac{di}{dt} + Ri = V = \text{constant}$$

given that current $i = 0$ at time $t = 0$.

Solution

Let us define

$$f(t) = \int_0^t x(t')\,dt'$$

It follows that $f(0) = 0$ and $df/dt = x(t)$. Now

$$\mathcal{L}\left\{\frac{df}{dt}\right\} = s\mathcal{L}\{f(t)\} - f(0)$$

Rewriting this in terms of x and F, we obtain

$$F(s) = s\mathcal{L}\left\{\int_0^t x(t')\,dt'\right\}$$

and division by s gives the required result. If we define $\mathcal{L}\{i(t)\} = I(s)$, then transforming the circuit equation leads to

$$LsI + RI = \frac{V}{s}$$

(after putting $i(0) = 0$) from which

$$I(s) = \frac{V}{Ls(s + R/L)}$$

This could, of course, be inverted using simple partial fractions, but noting that the factor s in the denominator corresponds to integration in the time domain, and that

$$\mathscr{L}^{-1}\left(\frac{1}{s + R/L}\right) = e^{-Rt/L}$$

we can put

$$i(t) = \frac{V}{L}\int_0^t e^{-Rt'/L}\,dt'$$

giving

$$i(t) = \frac{V}{R}[1 - e^{-Rt/L}] \qquad \bullet$$

Worked Example 1.8

A linear system is described by three simultaneous differential equations. There are two inputs $x(t)$ and three outputs $y(t)$, as illustrated in Fig. 1.4. All initial conditions are zero. In terms of Laplace transforms ($\mathscr{L}\{x_1(t)\} = \bar{x}_1(s)$, etc.) the system can be modelled in the form

$$\begin{bmatrix} \bar{y}_1 \\ \bar{y}_2 \\ \bar{y}_3 \end{bmatrix} = \begin{bmatrix} g_{11} & g_{12} \\ g_{21} & g_{22} \\ g_{31} & g_{32} \end{bmatrix} \begin{pmatrix} \bar{x}_1 \\ \bar{x}_2 \end{pmatrix}$$

in which the elements of the matrix G are also functions of s.

It is observed that when the input is

$$\begin{pmatrix} x_1 \\ x_2 \end{pmatrix} = \begin{pmatrix} \delta(t) \\ 0 \end{pmatrix}$$

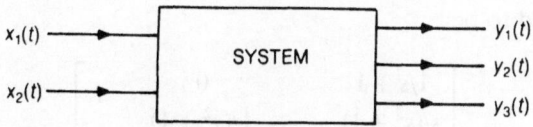

Fig. 1.4 A linear system.

the output is

$$
\begin{bmatrix} y_1 \\ y_2 \\ y_3 \end{bmatrix} = \begin{bmatrix} e^{-t} \\ \cos t \\ 1 \end{bmatrix}
$$

and that when the input is

$$
\begin{pmatrix} x_1 \\ x_2 \end{pmatrix} = \begin{pmatrix} 0 \\ \delta(t) \end{pmatrix}
$$

the output is

$$
\begin{bmatrix} y_1 \\ y_2 \\ y_3 \end{bmatrix} = \begin{bmatrix} 0 \\ \sin t \\ e^{-t}\cos t \end{bmatrix}
$$

Find the elements (g_{ij}) and obtain the output in terms of general inputs $x_1(t)$ and $x_2(t)$.

Solution

We may use the transform pairs of Table 1.2 and the first shift theorem (property L1 of Table 1.1) to replace the time-domain input–output data given above by the equivalent transforms. This leads to two sets of three equations as follows:

$$
\begin{bmatrix} 1/(s+1) \\ s/(s^2+1) \\ 1/s \end{bmatrix} = \begin{bmatrix} g_{11} & g_{12} \\ g_{21} & g_{22} \\ g_{31} & g_{32} \end{bmatrix} \begin{pmatrix} 1 \\ 0 \end{pmatrix}
$$

and

$$
\begin{bmatrix} 0 \\ 1/(s^2+1) \\ (s+1)/[(s+1)^2+1] \end{bmatrix} = \begin{bmatrix} g_{11} & g_{12} \\ g_{21} & g_{22} \\ g_{31} & g_{32} \end{bmatrix} \begin{pmatrix} 0 \\ 1 \end{pmatrix}
$$

From this the matrix G (known as the *transfer function* of the system) is identified to be

$$
\begin{bmatrix} 1/s+1 & 0 \\ s/(s^2+1) & 1/(s^2+1) \\ 1/s & (s+1)/[(s+1)^2+1] \end{bmatrix}
$$

and then

$$\begin{bmatrix} \bar{y}_1 \\ \bar{y}_2 \\ \bar{y}_3 \end{bmatrix} = G\begin{pmatrix} \bar{x}_1 \\ \bar{x}_2 \end{pmatrix}$$

provides the transform of the output for any input. In particular

$$\bar{y}_1 = \frac{1}{s+1}\bar{x}_1$$

$$\bar{y}_2 = \frac{s}{s^2+1}\bar{x}_1 + \frac{1}{s^2+1}\bar{x}_2$$

and

$$\bar{y}_3 = \frac{1}{s}\bar{x}_1 + \frac{(s+1)}{[(s+1)^2+1]}\bar{x}_2$$

If the inputs $x_1(t)$ and $x_2(t)$ are not specified, we may express the outputs in terms of convolution integrals, using property L6 of Table 1.1. This gives the results

$$y_1(t) = \int_0^t e^{-t'}x_1(t-t')\,dt = \int_0^t e^{-(t-t')}x_1(t')\,dt'$$

$$y_2(t) = \int_0^t \{(\cos t')x_1(t-t') + (\sin t')x_2(t-t')\}\,dt'$$

$$= \int_0^t \{[\cos(t-t')]x_1(t') + [\sin(t-t')]x_2(t')\}\,dt'$$

$$y_3(t) = \int_0^t \{x_1(t-t') + e^{-t'}(\cos t')x_2(t-t')\}\,dt'$$

$$= \int_0^t \{x_1(t') + e^{-(t-t')}[\cos(t-t')]x_2(t')\}\,dt'$$

We note that the symmetry of the convolution means that each output can be expressed in two ways. In practice one integral may be easier to evaluate than the other – but the results, of course, are the same in each case. ●

1.9 TRANSFER FUNCTIONS AND CONVOLUTION

In Worked Example 1.8 we referred to the 'transfer function' of a system when the latter is described in the transform domain. In the

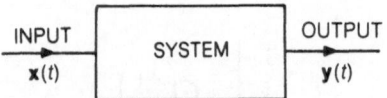

Fig. 1.5 A general linear system.

time domain, if the system is described by differential equations, then the input–output diagram is as in Fig. 1.5, which represents equations of the form

$$M(\mathrm{D}) \cdot \mathbf{y}(t) = \mathbf{x}(t)$$

or

$$\mathbf{y}(t) = [M(\mathrm{D})]^{-1} \cdot \mathbf{x}(t)$$

in which D is the differential operator d/dt and $M(\mathrm{D})$ is the matrix whose rows are the (polynomial) differential operators of the differential equations appearing in the system description. (We have used vector notation $\mathbf{x}(t)$ and $\mathbf{y}(t)$ to indicate that there might be several inputs and outputs.)

In the transform domain, we have the system description illustrated in Fig. 1.6 (meaning $Y(f) = H(f)X(f)$, whatever transform is applied, f being the transform parameter.)

Clearly, properties which relate convolution integrals in the time domain to products in the transform domain (and conversely) are very important, whether we are wanting to identify the transfer function $H(f)$ or to predict the response of a system to known inputs when $H(f)$ is known. The latter is sometimes referred to as the *pulse* transfer function because, as illustrated in Worked Example 1.8, if the input is a single pulse then $Y(f) = H(f)$ gives the unit pulse response for the particular system characterized by $H(f)$. In Chapter 2 we shall examine convolution and multiplication properties in more detail in the context of the Fourier transform and, more generally, consider how the evaluation of convolution integrals can be assisted by graphical considerations – especially when dealing with functions defined in a piecewise-continuous manner. Subsequently we shall

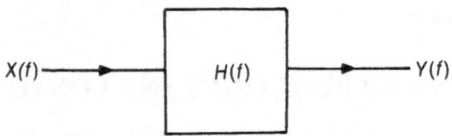

Fig. 1.6 Transform of a linear system.

consider convolution in the context of *discrete* signals and *discrete* transforms, as required for digital computation.

SUMMARY

In this chapter we have attempted to review and extend the reader's background knowledge of various ways of representing a function by a series, with particular reference to the information (concerning harmonic content) about a signal which is conveyed by such a representation. We have further considered various types of transform which might be applied to a function defined on a range which is not finite, and indicated some of the advantages to be gained by describing a system in the transform domain rather than in the frequency domain.

PROBLEMS

1.1 Show that the Fourier exponential series for

$$x(t) = \sin(\alpha t), \quad -l < t < l$$

if $\alpha \neq N\pi/l$, where N is an integer, is

$$x(t) \sim -j\pi \sin(\alpha l) \sum_{n=-\infty}^{\infty} \frac{(-1)^n n}{\alpha^2 l^2 - n^2 \pi^2} e^{jn\pi t/l}$$

If $\alpha = N\pi/l$, show that the coefficient $X_N = -j/2$.

1.2 A pulse train of period $T = 1$ and amplitude 1 is described by the equation

$$x(t) = \sum_{m=-\infty}^{\infty} [H(t-m) - H(t-m-\tfrac{1}{4})]$$

Show that the coefficients in the complex Fourier series are given by

$$X_n = \frac{e^{-jn\pi/4} \sin(n\pi/4)}{n\pi}$$

1.3 Show that the Fourier transform of the function

$$x(t) = \begin{cases} 1 & |t| < a \\ 0 & |t| > a \end{cases}$$

is

$$X(f) = \frac{\sin(2\pi f a)}{\pi f}$$

1.4   Show that the (one-sided) Fourier transform of $x(t) = e^{-t} \sin t$
      can be written

$$X(w) = \frac{(2 - w^2) - 2jw}{4 + w^4}$$

Confirm that this result can be obtained by appropriate modification of the Laplace transform of $e^{-t} \sin t$.

1.5   If

$$x(t) = x(t + T) = x(t + nT)$$

then by writing

$$\int_0^\infty x(t')e^{-st'}\, dt' = \int_0^T x(t')e^{-st'}\, dt' + \int_T^\infty x(t')e^{-st'}\, dt'$$

show that the Laplace transform of a periodic function is as stated in property L5 of Table 1.1.

1.6   Given $x_1(t) = e^{at}$ and $x_2(t) = \sin(at)$, show that

$$\int_0^t x_1(t - t')x_2(t')\, dt' = \frac{1}{2a}[e^{at} - \sin(at) - \cos(at)].$$

Transform this result using the Laplace transform Table 1.2 and verify that the convolution property L6 of Table 1.1 is satisfied.

1.7   Use Laplace transform methods to show that the solution of the equation

$$L\frac{d^2q}{dt^2} + \frac{1}{C}q = E_0\delta(t - T)$$

where $q = dq/dt = 0$ at time $t = 0$, is

$$q(t) = E_0\sqrt{\frac{C}{L}}\left(\sin\frac{t - T}{\sqrt{LC}}\right)H(t - T)$$

1.8   The function $J_0(t)$ has a Laplace transform $F(s) = 1/\sqrt{s^2 + 1}$. Use the convolution property to deduce that

$$\int_0^t J_0(t')J_0(t - t')\, dt' = \sin t$$

(Note that $J_0(t)$ is the *Bessel function* of order zero. Bessel functions appear, for example, in the description of radio antennae subject to horizontal deflection. More generally, if a partial differential equation describing a system is written in terms of plane or cylindrical polar coordinates, Bessel functions appear in the solution.)

1.9 A damped vibrating system is described by the equation

$$\frac{d^2y}{dt^2} + 6\frac{dy}{dt} + 10y = x(t)$$

and $y(0) = 1$, $y'(0) = 0$ are the initial displacement and velocity. The input $x(t)$ is not defined. Use Laplace transform methods (including the convolution property) to show that the output may be written

$$y(t) = e^{-3t}(\cos t + 3\sin t) + \int_0^t e^{-3t'}x(t - t')\sin t'\, dt'$$

1.10 Letting $F_1(s) = \mathscr{L}\{x_1(t)\}$ and $F_2(s) = \mathscr{L}\{x_2(t)\}$, transform the simultaneous differential equations

$$\frac{dx_1}{dt} + x_1 - x_2 = 6[1 - H(t - 1)]$$

$$\frac{dx_2}{dt} + x_1 + x_2 = 6H(t - 1)$$

given that $x_1(0) = 3$ and $x_2(0) = -2$. Using any appropriate methods and inversion properties, obtain the results

$$x_1(t) = e^{-t}\sin t + 3 - 6H(t - 1)e^{-(t-1)}\sin(t - 1)$$
$$x_2(t) = e^{-t}\cos t - 3 - 6H(t - 1)[e^{-(t-1)}\cos(t - 1) - 1]$$

Confirm that the initial conditions are satisfied. What happens as $t \to \infty$?

1.11 (i) Use Laplace transform properties and pairs to show that

$$\mathscr{L}\left\{\int_0^t x(t - u)e^{-2u}\cos 3u\, du\right\} = \frac{(s + 2)F(s)}{s^2 + 4s + 13}$$

(ii) It was shown in Worked Example 1.7 that

$$\mathscr{L}\left\{\int_0^t x(t')\, dt'\right\} = \frac{F(s)}{s}$$

Show that the integral equation

$$x(t) + 2\int_0^t x(t - u)e^{-2u}\cos 3u\, du + 2\int_0^t x(u)\, du = e^{-5t}$$

has solution

$$x = -\frac{6}{5}e^{-2t} + \frac{15}{4}e^{-5t} + \frac{e^{-3t}}{20}(12\sin 2t - 31\cos 2t)$$

1.12 A light horizontal beam of uniform cross-section is clamped at its ends $x = 0$ and $x = l$ and carries a concentrated load W at a point $x = a$, where $0 < a < l$. The vertical deflection is $y(x)$ and the system is described by the equations

$$\frac{d^4y}{dx^4} = \frac{W}{EI}\delta(x - a) \qquad y = \frac{dy}{dx} = 0 \text{ at } x = 0 \text{ and } x = l$$

in which E and I are physical constants.

Use Laplace transform methods to show that the deflection distribution takes the form

$$y(x) = \frac{W}{EI}\left[H(x - a)\frac{(x - a)^3}{3!} + A_1\frac{x^3}{3!} + A_2\frac{x^2}{2!}\right]$$

where A_1 and A_2 can be found from the equations

$$0 = \frac{W}{EI}\left[\frac{(l - a)^3}{3!} + \frac{A_1 l^3}{3!} + \frac{A_2 l^2}{2!}\right]$$

$$0 = \frac{W}{EI}\left[\frac{(l - a)^2}{2!} + \frac{A_1 l^2}{2!} + A_2 l\right]$$

2

The Fourier transform. Convolution of analogue signals

Having recognized that harmonic analysis of a signal is an extremely useful facility, and that transform techniques offer a powerful means of identifying the 'transfer' characteritics of a system, those methods will now be examined in greater depth.

In practical situations commercial apparatus is available which can immediately convert online experimental data into a digital or graphical output description of a frequency spectrum. Applications are found in engine design, naval architecture, and the analysis of the behaviour of tracked vehicles, for example. Factors affecting system performance might be internal (such as engine noise consequent on throttle-setting, or 'knocking' in heat-exchanger tubes due to cavitation in the upline steam generator), or external (such as received vibrations due to travelling over uneven terrain at particular speeds, or oceanic wave-patterns). The processing of discrete data, as required in such contexts, is more clearly understood by first considering the treatment of *continuous* signals. The idea of *convolution* will emerge naturally, as a consequence of choosing to treat a system description in the frequency domain rather than in the time domain, or conversely.

Accordingly, in this chapter we propose to establish some key properties of the Fourier transform, and in particular to show how the symmetry property of the transform facilitates the establishment of other transform properties and pairs when derivation from the defining integral presents difficulties. After proving (*inter alia*) the convolution property for the Fourier transform, we shall look at convolution integrals more generally to show how graphical illustration of the integrand factors can assist with the integration process – in particular to determine exactly *what* is to be integrated in different intervals of the range of integration.

Various commonly occurring functions will be defined in Sections 2.4 and 2.6.

The material of Chapter 2 relates to functions which are continuous (or piecewise continuous) functions of time or of the transform parameter, in order to provide the background for the consideration of discrete signals and discrete transforms which forms the subject matter of the rest of the book.

2.1 DUALITY

In Section 1.5 we established the Fourier transform pair

$$X(f) = \int_{-\infty}^{\infty} x(t')e^{-j2\pi ft'}\,dt' = \mathcal{F}\{x(t)\} \tag{2.1}$$

$$x(t) = \int_{-\infty}^{\infty} X(f)e^{j2\pi ft}\,df = \mathcal{F}^{-1}\{X(f)\} \tag{2.2}$$

If, in (2.2), we exchange f and t we obtain

$$x(f) = \int_{-\infty}^{\infty} X(t)e^{j2\pi ft}\,dt$$

In this, if f is replaced by $-f$, the result is the equation

$$x(-f) = \int_{-\infty}^{\infty} X(t)e^{-j2\pi ft}\,dt \tag{2.3}$$

Comparing this with (2.1), we see that (2.3) can be regarded as giving the transform of a time-domain signal $X(t)$. In other words, if the transform $X(f)$ of $x(t)$ is known, then the transform of $X(t)$ is given by

$$\mathcal{F}\{X(t)\} = x(-f) \tag{2.4}$$

This is known as the *duality* property of the Fourier transform, and appears as property F1 in Table 2.1. (There is no such equivalent property of the Laplace transform.)

We also note from the similarity of (2.1) and (2.2) that any computational algorithm developed to obtain the transform of a signal can, with appropriate sign changes or conjugation, be used to recover the time-domain signal from a known transform.

Sign change is also the subject of the *reversal* property (F2 of Table 2.1), which it is appropriate to include in this section. Changing the

Table 2.1 Fourier transform properties, $X(f) = \mathscr{F}\{x(t)\}$

Property	Time domain	Transform domain
F1 Duality	$X(t)$	$x(-f)$
F2 Reversal	$x(-t)$	$X(-f)$
F3 Scaling	$x(at)$	$\dfrac{1}{\|a\|}X\left(\dfrac{f}{a}\right)$
F4 Delay	$x(t-t_0)$	$e^{-j2\pi f t_0}X(f)$
F5 Translation (rotation)	$e^{j2\pi f_0 t}x(t)$	$X(f-f_0)$
F6 Differentiation in time	(i) $\dfrac{dx}{dt}$	$(j2\pi f)X(f)$
	(ii) $\dfrac{d^n x}{dt^n}$	$(j2\pi f)^n X(f)$
F7 Convolution	$\displaystyle\int_{-\infty}^{\infty} x_1(t-t')x_2(t')\,dt'$	$X_1(f)X_2(f)$
	$=\displaystyle\int_{-\infty}^{\infty} x_1(t')x_2(t-t')\,dt'$	
F8 Products	$x_1(t)x_2(t)$	$\displaystyle\int_{-\infty}^{\infty} X_1(f-f')X_2(f')\,df'$
		$=\displaystyle\int_{-\infty}^{\infty} X_1(f')X_2(f-f')\,df'$

sign of t in $x(t)$, from (2.1) the transform of $x(-t)$ is defined by

$$\mathscr{F}\{x(-t)\} = \int_{-\infty}^{\infty} x(-t')e^{-j2\pi f t'}\,dt'$$

Substituting $t' = -\tau$ gives

$$\mathscr{F}\{x(-t)\} = \int_{-\infty}^{\infty} x(\tau)e^{j2\pi f \tau}\,d\tau$$

which we can regard as equivalent to (2.1) in which f has been replaced by $-f$. Hence

$$\mathscr{F}\{x(-t)\} = X(-f) \qquad (2.5)$$

This result is used in the next worked example, which is designed to illustrate the value of the duality property in a case when the defining integral of the transform does not exist as such.

Worked Example 2.1

Obtain the Fourier transform of $x(t) = \delta(t - t_0)$ and deduce the Fourier transform of $e^{j2\pi f_0 t}$.

Solution

By definition

$$\mathscr{F}\{\delta(t - t_0)\} = \int_{-\infty}^{\infty} \delta(t - t_0)e^{-j2\pi ft}\,dt$$

Using the sifting property of the impulse function (which was given in the footnote to Table 1.2), we have

$$\mathscr{F}\{\delta(t - t_0)\} = e^{-j2\pi f t_0} = X(f)$$

From the duality property, (2.4), we know that the transform of $X(t)$ is $x(-f)$, and from the reversal property, (2.5), we know that the transform of $X(-t)$ is then $x(-[-f]) = x(f)$. It follows that the transform of $X(-t) = e^{j2\pi f_0 t}$ is $\delta(f - f_0)$.

This result could be obtained using other properties of the Fourier transform, but we note that direct use of the defining integral (2.1) would require the 'evaluation' of

$$\mathscr{F}\{e^{j2\pi f_0 t}\} = \int_{-\infty}^{\infty} e^{j2\pi(f_0 - f)t'}\,dt'$$

and even a 'transform in the limit' approach would present problems. (This approach will be discussed in Section 2.4.) ●

2.2 FURTHER PROPERTIES OF THE FOURIER TRANSFORM

The reversal property can be regarded as a special case (with $a = -1$) of the *scaling* property (F3 of Table 2.1) which states that

$$\mathscr{F}\{x(at)\} = \frac{1}{|a|}X\left(\frac{f}{a}\right) \tag{2.6}$$

(The proof of this is required at the end of the chapter in Problem 2.1.)

To prove the *delay* property (F4), we start with the definition

$$\mathscr{F}\{x(t-t_0)\} = \int_{-\infty}^{\infty} x(t'-t_0)e^{-j2\pi ft'}\,dt'$$

$$= \int_{-\infty}^{\infty} x(\tau)e^{-j2\pi f(\tau+t_0)}\,d\tau$$

(on substituting $t' = t_0 + \tau$)

$$= e^{-j2\pi ft_0} \int_{-\infty}^{\infty} x(\tau)e^{-j2\pi f\tau}\,d\tau$$

Using the definition of the Fourier transform again, we have the result

$$\mathscr{F}\{x(t-t_0)\} = e^{-j2\pi ft_0}X(f) \tag{2.7}$$

The *translation* property (F5) can also be proved from the defining integral, but we note that the result

$$\mathscr{F}\{e^{j2\pi f_0 t}x(t)\} = X(f-f_0) \tag{2.8}$$

is a direct consequence of (2.7) if we invoke the duality property.

Properties F4 and F5 can be regarded as the equivalent of the Laplace transform shift theorems, in that a shift in the time domain results in the appearance of an exponential factor in the frequency (or transform) domain, and conversely. The exponential factor can be regarded as effecting a 'rotation', in the sense that we can regard $e^{j\alpha}$ as a vector operator which rotates a vector through an angle α if α is a real quantity. (Consider the vector representing a complex number $z = re^{j\theta}$. Multiplication of z by $e^{j\alpha}$ increases the argument to $\theta + \alpha$.)

The *differentiation* property (F6) of the Fourier transform can be compared with that for the Laplace transform (L3 of Table 1.1), with the 'correspondence' between parameters s, jw, $j2\pi f$ in mind. The initial values present in L3, however, have no counterpart in the result for the two-sided Fourier transform.

Worked Example 2.2 (Property F6)

Show that if $\mathscr{F}\{x(t)\} = X(f)$ then

$$\mathscr{F}\{x^{(n)}(t)\} = (j2\pi f)^n X(f) \tag{2.9}$$

Solution

By definition,

$$\mathcal{F}\{x^{(n)}(t)\} = \int_{-\infty}^{\infty} \frac{d^n x}{dt^n} e^{-j2\pi ft'} \, dt'$$

$$= \left[\frac{d^{n-1}x}{dt^{n-1}} e^{-j2\pi ft'} \right]_{-\infty}^{\infty} + j2\pi f \int_{-\infty}^{\infty} \frac{d^{n-1}}{dt^{n-1}} e^{-j2\pi ft'} \, dt'$$

integrating by parts. At this stage, anticipating the conditions stated in Section 2.3, we assume that

$$\lim_{t \to \pm \infty} \frac{d^{n-1}x}{dt^{n-1}} = 0$$

and so

$$\mathcal{F}\{x^{(n)}(t)\} = (j2\pi f)\mathcal{F}\{x^{(n-1)}(t)\}$$

This result can be viewed as a reduction, or recurrence, equation to be used inductively in that

$$\mathcal{F}\{x^{(n-1)}(t)\} = (j2\pi f)\mathcal{F}\{x^{(n-2)}(t)\}$$

from which

$$\mathcal{F}\{x^{(n)}(t)\} = (j2\pi f)^2 \mathcal{F}\{x^{(n-2)}(t)\}$$

and so on. After repeated use of the same equation we will obtain the result

$$\mathcal{F}\{x^{(n)}(t)\} = (j2\pi f)^k \mathcal{F}\{x^{(n-k)}(t)\}$$

and ultimately, when $k = n$,

$$\mathcal{F}\{x^{(n)}(t)\} = (j2\pi f)^n \mathcal{F}\{x(t)\} = (j2\pi f)^n X(f).$$

This verifies (2.9).   ●

The convolution integral of two functions will now be defined (more generally than in the case given in property L6 of Table 1.1) as

$$\int_{-\infty}^{\infty} x_1(t - t')x_2(t') \, dt' = \int_{-\infty}^{\infty} x_1(t')x_2(t - t') \, dt'$$

(That these integrals are equal can be verified by a change of variable.) The usual notation for the convolution of two signals is

$$x(t) = x_1(t) * x_2(t)$$

meaning either of the above integral expressions.

Worked Example 2.3

Show that

$$\mathscr{F}\{x_1 * x_2\} = X_1(f)X_2(f) \qquad (2.10)$$

Solution

Let us begin with the version of the convolution which reads

$$x(t) = x_1 * x_2 = \int_{-\infty}^{\infty} x_1(t - t')x_2(t')\,dt'$$

We can quote the delay property (F4) to note that

$$\mathscr{F}\{x_1(t - t')\} = e^{-j2\pi ft'}X_1(f)$$

which has inverse

$$x_1(t - t') = \int_{-\infty}^{\infty} e^{j2\pi ft}\{e^{-j2\pi ft'}X_1(f)\}\,df$$

(The above equation incorporates both the defining integral of the transform and its inverse.)

From our first expression for the convolution

$$x(t) = x_1(t) * x_2(t)$$

we now have an equation

$$x(t) = \int_{-\infty}^{\infty} x_2(t')\left[\int_{-\infty}^{\infty} X_1(f)e^{j2\pi f(t-t')}\,df\right]dt' \qquad (2.11)$$

In (2.11) it is implied that the integration with respect to f precedes that with respect to t'. If we change the order of integration, (2.11) can be written

$$x(t) = \int_{-\infty}^{\infty} X_1(f)e^{j2\pi ft}\left[\int_{-\infty}^{\infty} x_2(t')e^{-j2\pi ft'}\,dt'\right]df$$

There appears in this the definition of the transform of $x_2(t)$, and so

$$x(t) = \int_{-\infty}^{\infty} X_1(f)X_2(f)e^{j2\pi ft}\,df$$

This is, in effect, the inversion formula for

$$\mathscr{F}^{-1}\{X_1(f)X_2(f)\} = x_1(t) * x_2(t)$$

hence

$$\mathscr{F}\{x_1(t)*x_2(t)\} = X_1(f)X_2(f) \qquad \bullet$$

The final property in Table 2.1 (F8) is the dual of the last result:

$$\mathscr{F}\{x_1(t)x_2(t)\} = \int_{-\infty}^{\infty} X_1(f-f')X_2(f')\mathrm{d}f'$$

$$= \int_{-\infty}^{\infty} X_1(f')X_2(f-f')\mathrm{d}f'$$

$$= X_1(f)*X_2(f).$$

The relation between multiplication in the time domain and convolution in the transform domain, and conversely, is extremely important, particularly when we proceed to work with discrete sampled signals. (This will emerge in Section 3.6.)

2.3 COMPARISON WITH THE LAPLACE TRANSFORM, AND THE EXISTENCE OF THE FOURIER TRANSFORM

In Section 1.8 we referred to the fact that if a function $x(t)$ is piecewise continuous and exponentially bounded on $t \geqslant 0$, then its Laplace transform exists.

For the integral defining the two-sided Fourier transform

$$\int_{-\infty}^{\infty} x(t')\mathrm{e}^{-j2\pi ft'}\,\mathrm{d}t'$$

to exist, conditions are stronger. We require that

$$\int_{-\infty}^{\infty} |x(t')|\mathrm{d}t'$$

exist, which includes the requirement that

$$\lim_{t\to\pm\infty} x(t) = 0$$

The one-sided Fourier transform might exist when the two-sided transform does not. In Worked Example 1.4 we obtained the one-sided transform of $x(t) = \mathrm{e}^{-t}\cos t$, which does *not* have a two-sided transform because on $t < 0$ the function exhibits oscillatory behaviour of increasing amplitude as $t \to -\infty$. The one-sided transform, however, could in that example be deduced from

$$\mathcal{L}\{e^{-t}\cos t\} = \frac{(s+1)}{(s+1)^2 + 1}$$

On replacing s by jw (following the comparison described in Section 1.7), we obtain

$$\mathscr{F}\{e^{-t}\cos t\} = \frac{jw+1}{(jw+1)^2+1} \cdot \frac{(-jw+1)^2+1}{(-jw+1)^2+1} = \frac{(2.+w^2)-jw^3}{4+w^4}$$

which was the result obtained by direct evaluation of

$$\int_0^\infty e^{-t'}\cos t' e^{-jwt'}\,dt'$$

Worked Example 2.4

Find the Fourier transform of

$$x(t) = e^{-at}H(t) \qquad a>0$$

Solution

Since $H(t) = 0$ on $t < 0$ this 'two-sided' transform is in effect a one-sided transform. Either by integrating

$$\int_0^\infty e^{-at'}e^{-j2\pi ft'}\,dt'$$

directly, or by quoting the result

$$\mathcal{L}\{e^{-at}\} = \frac{1}{s+a}$$

and replacing s by $jw = j2\pi f$, we obtain

$$\mathscr{F}\{e^{-at}H(t)\} = \frac{1}{j2\pi f + a} \qquad\qquad \bullet$$

2.4 TRANSFORMING USING A LIMIT PROCESS

To establish whether or not a function $x(t)$ has a Fourier transform, it might be of assistance to consider first the existence of the transform of a modified function, such as

$$x_a(t) = x(t)e^{-a|t|} \qquad a > 0$$

so that if

$$\int_{-\infty}^{\infty} x_a(t')e^{-j2\pi ft'}\, dt' = X_a(f)$$

exists we can then examine

$$\lim_{a \to 0} X_a(f)$$

This type of approach is illustrated in the next worked example, which concerns the signum function defined as follows:

$$\operatorname{sgn} t = \begin{cases} 1 & \text{if } t > 0 \\ -1 & \text{if } t < 0 \end{cases}.$$

Worked Example 2.5

By considering the function

$$x_a(t) = e^{-at}H(t) - e^{at}H(-t) \qquad a > 0$$

find the Fourier transform of sgn t. Use the duality property to deduce the transform of $1/t$.

Solution

Noting that $H(-t)$ is the 'folded' step function equal to 1 on $t < 0$ and zero on $t > 0$, the graph of the function $x_a(t)$ is as shown in Fig. 2.1, and $\lim_{a \to 0} x_a(t) = \operatorname{sgn} t$. From the defining integral (2.1) we

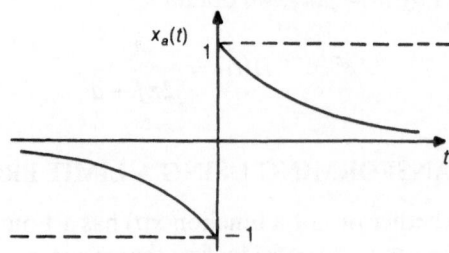

Fig. 2.1 The pre-limit function, $x_a(t)$.

have

$$\mathscr{F}\{x_a(t)\} = \int_{-\infty}^{0} -e^{at'}e^{-j2\pi ft'}\,dt' + \int_{0}^{\infty} e^{-at'}e^{-j2\pi ft'}\,dt'$$

$$= \left[\frac{-e^{(a-j2\pi f)t'}}{a-j2\pi f}\right]_{-\infty}^{0} + \left[\frac{-e^{-(a+j2\pi ft')}}{a+j2\pi f}\right]_{0}^{\infty}$$

$$= \frac{-j4\pi f}{a^2 + 4\pi^2 f^2} = X_a(f)$$

Letting $a \to 0$, it follows that

$$\mathscr{F}\{\operatorname{sgn} t\} = \frac{-j}{\pi f} = X(f) \qquad (2.12)$$

This can be written $\mathscr{F}\{j\pi \operatorname{sgn} t\} = 1/f$ and so, from the duality property $\mathscr{F}\{X(t)\} = x(-f)$, we have the second transform pair

$$\mathscr{F}\left\{\frac{1}{t}\right\} = -j\pi \operatorname{sgn} f. \qquad (2.13)$$

(These results, and other pairs already obtained or about to be obtained, are included in Table 2.2 below.)　　　　　　●

This example again illustrates the power of the duality property of the Fourier transform, as – whether or not a limit process is invoked – the dual result (2.13) is not readily obtained from consideration of the integral

$$\int_{-\infty}^{\infty} \frac{1}{t'} e^{-j2\pi ft'}\,dt'$$

Indeed, in many cases where the existence of the defining integral presents difficulties, these might be circumvented by use of the properties of the Fourier transform. Otherwise we might conclude that even some quite simple functions have no transform, strictly speaking. We shall also see, in the context of periodic functions in particular, that the prior representation of the function by a Fourier series can be helpful in establishing an expression for a Fourier transform.

2.5 TRANSFORMATION AND INVERSION USING DUALITY AND OTHER PROPERTIES

Although most properties are proved using the defining integral of the Fourier transform (or any other transform), as has just been stated we are more likely to make progress using 'properties' rather than the defining integral (2.1) and the inversion integral (2.2) when deriving transform pairs.

Worked Example 2.6

Show that

$$x(t) = e^{j2\pi t} e^{-(t-2)} H(t-2)$$

and

$$X(f) = \frac{e^{-j4\pi f}}{1 + j2\pi(f-1)}$$

constitute a Fourier transform pair

(i) by transforming $x(t)$
(ii) by inverting $X(f)$

Solution

(i) We could use the defining integral (2.1) to evaluate

$$\mathscr{F}\{x(t)\} = \int_{-\infty}^{\infty} e^{j2\pi t'} e^{-(t'-2)} H(t'-2) e^{-j2\pi f t'} \, dt'$$

$$= \int_{2}^{\infty} e^{j2\pi t'} e^{-(t'-2)} e^{-j2\pi f t'} \, dt'$$

directly, with little difficulty. However, from Worked Example 2.4 it has already been established that

$$\mathscr{F}\{e^{-t}H(t)\} = \frac{1}{j2\pi f + 1}$$

From the delay property (F4 of Table 2.1) it follows that (putting $t_0 = 2$)

$$\mathscr{F}\{e^{-(t-2)}H(t-2)\} = e^{-j4\pi f} \frac{1}{j2\pi f + 1}$$

and using the rotation property (F5 of Table 2.1) we then have (putting $f_0 = 1$)

$$\mathscr{F}\{e^{j2\pi t} e^{-(t-2)} H(t-2)\} = e^{-j4\pi(f-1)} \frac{1}{j2\pi(f-1)+1} \quad (2.14)$$

which is $X(f)$ (noting that $e^{j4\pi} = 1$).

(ii) To use (2.2) and write

$$x(t) = \int_{-\infty}^{\infty} \frac{e^{-j4\pi f} e^{j2\pi ft}}{1 + j2\pi(f-1)} \, df$$

would *not* be a promising start to an inversion process.
Instead, we begin with (2.14):

$$X(f) = e^{-j4\pi(f-1)} \frac{1}{1 + j2\pi(f-1)}$$

and use the same properties as in (i) but in reverse order. Namely, $X(f)$ has an inverse

$$e^{j2\pi t} x_1(t)$$

where $x_1(t)$ has a transform $e^{-j4\pi f}/(1 + j2\pi f)$ (using the rotation property F5). This, in turn, is the transform of $x_1(t-2)$ (using the delay property F4), and this next identifies $x_1(t)$ as the 'basic' function whose transform is $1/(1 + j2\pi f)$, which is $e^{-t}H(t)$.

Hence the inverse of (2.14) is

$$x(t) = e^{j2\pi t} x_1(t-2)$$
$$= e^{j2\pi t} e^{-(t-2)} H(t-2)$$

(It might help the reader to compare this with the steps taken to invert a known Laplace transform when using both first and second shift theorems in connection with one, basic, tabulated transform function.) ●

2.6 SOME FREQUENTLY OCCURRING FUNCTIONS AND THEIR TRANSFORMS

We have already rehearsed the definitions of the unit step function $H(t-a)$ and of the impulse function $\delta(t-a)$. The signum function has also been introduced (see Section 2.4).

The *on/off* property of the H-function provides us with a 'multiplier'

$$H(t-a) - H(t-b)$$

as illustrated in Fig. 2.2. This multiplier is equal to 1 if $a < t < b$ but is zero outside that range, and it therefore provides us with a factor which we can apply to a signal which is non-zero over a specified finite range. For example, we can graph $[H(t-\pi) - H(t-2\pi)]\sin t$ in the form shown in Fig. 2.3.

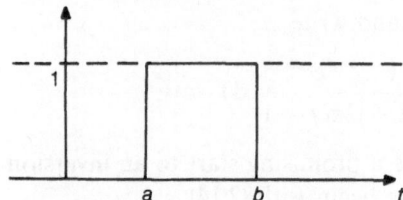

Fig. 2.2 The rectangle function.

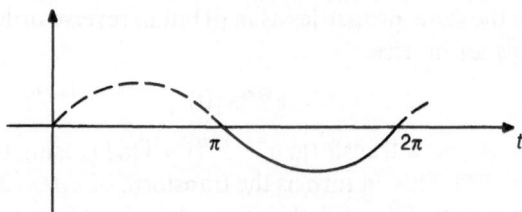

Fig. 2.3 A single half-cycle of the sine wave.

It is often the case that we need to consider *finite-duration* signals, and to that end we first define

$$\Pi\left(\frac{t - t_0}{B}\right)$$

to be a function of amplitude 1, centred at t_0, width B, as illustrated in Fig. 2.4. This could be written in terms of H-functions, but provides the zero/one factors more economically. (The area enclosed is equal to B.)

The function

$$x(t) = \begin{cases} 1 & |t| < a \\ 0 & |t| > a \end{cases}$$

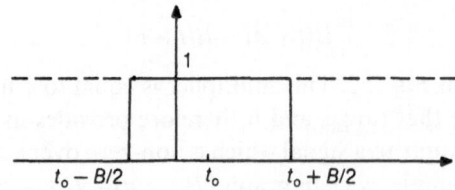

Fig. 2.4 The pi (or pulse) function, $\Pi[(t - t_0)/B]$.

was the subject of Problem 1.3, and can be written as

$$x(t) = \Pi\left(\frac{t}{2a}\right)$$

The transform of $x(t)$ was required in the form $\sin(2\pi fa)/\pi f$, and in terms of the sinc function defined below, this can be written

$$X(f) = 2a \operatorname{sinc}(2fa)$$

Next, we define

$$\Lambda\left(\frac{t - t_0}{B}\right)$$

to be the triangle function centred at t_0, width $2B$ (so that the area under the 'curve' is B, as for Π), shown in Fig. 2.5.

In Worked Example 1.3 we showed that the Fourier transform of

$$x(t) = \begin{cases} t + 1 & -1 < t < 0 \\ 1 - t & 0 < t < 1 \\ 0 & |t| > 1 \end{cases}$$

is

$$X(f) = \left(\frac{\sin(\pi f)}{\pi f}\right)^2$$

These can be written as

$$x(t) = \Lambda(t)$$

and

$$X(f) = \operatorname{sinc}^2 f$$

respectively (again anticipating the following definition of the sinc function).

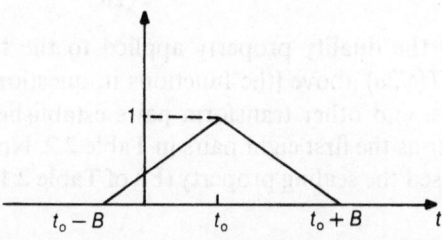

Fig. 2.5 The lambda (or triangle) function, $\Lambda[(t - t_0)/B]$.

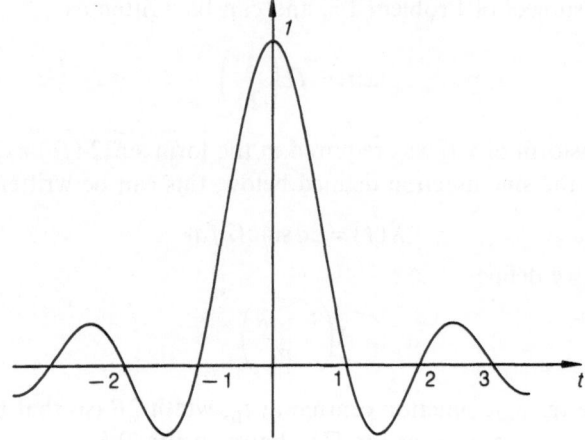

Fig. 2.6 The sinc function, sinc t.

Next, we define

$$\text{sinc}\, t = \frac{\sin(\pi t)}{\pi t} \tag{2.15}$$

This function, as is seen in Fig. 2.6, has all the zeros of $\sin(\pi t)$ except for $t = 0$ where $\sin(\pi t)/\pi t$ is indeterminate. If we define $\text{sinc}\, t = 1$ at $t = 0$, the function is continuous because using either the 'small angle' approximation or L'Hôpital's rule, we obtain

$$\lim_{t \to 0}\left(\frac{\sin(\pi t)}{\pi t}\right) = 1$$

Also, $\sin(\pi t)/\pi t$ tends to zero as $|t| \to \infty$.

We observe that the result

$$\mathscr{F}\{2a\,\text{sinc}\,2at\} = \Pi\left(\frac{f}{2a}\right)$$

is provided by the duality property applied to the transform pair explicitly for $\Pi(t/2a)$ above (the functions in question being even).

These results, and other transform pairs established in previous sections, appear as the first eight pairs in Table 2.2. Note that in pair FT3 we have used the scaling property (F3 of Table 2.1) to generalize the result

$$\mathscr{F}\{\Lambda(t)\} = \text{sinc}^2 f$$

given above. Similarly, we could use the delay property to transform Π and Λ functions centred at $t_0 \neq 0$.

A function related to the sinc function might be mentioned at this stage, although we do not need to consider its transform here. Define the sine integral function

$$\text{Si}(t) = \int_0^t \frac{\sin t'}{t'}\,dt' = \int_{-t}^0 \frac{\sin t'}{t'}\,dt' \tag{2.16}$$

This function appears, for example, if we are 'viewing' part of a transform function through a rectangular 'window' and so are considering (for example) a product such as $X(f)\Pi(f/B)$ in the transform domain. (Windowing has the effect of truncating the transform $X(f)$, because $\Pi(f/b)$ is zero outside the range $-B/2 < f < +B/2$.) Remember that, on inversion, this implies convolution in the time domain (property F7 of Table 2.1), and that Π-functions in the frequency domain are transforms of sinc functions in the time domain. It follows that if, in the time domain, $x(t)$ is itself a Π-function (or

Table 2.2 Fourier transform pairs

	$x(t)$	$X(f)$
FT1	$\Pi\left(\dfrac{t}{B}\right)$	$B\,\text{sinc}(Bf)$
FT2	$A\,\text{sinc}(At)$	$\Pi\left(\dfrac{f}{A}\right)$
FT3	$\Lambda\left(\dfrac{t}{B}\right)$	$B\,\text{sinc}^2(Bf)$
FT4	$\delta(t-t_0)$	$e^{-j2\pi f t_0}$
FT5	$e^{j2\pi f_0 t}$	$\delta(f-f_0)$
FT6	$e^{-at}H(t),\ a>0$	$\dfrac{1}{j2\pi f + a}$
FT7	$\text{sgn}\,t$	$\dfrac{1}{j\pi f}$
FT8	$\dfrac{1}{t}$	$-j\pi\,\text{sgn}\,f$
FT9	$e^{-a|t|},\ a>0$	$\dfrac{2a}{(2\pi f)^2 + a^2}$
FT10	$\displaystyle\sum_{m=-\infty}^{\infty}\delta(t-mT_s)$	$f_s\displaystyle\sum_{m=-\infty}^{\infty}\delta(f-mf_s),\ f_s=\dfrac{1}{T_s}$

a combination of *H*-functions), the convolution will involve integrals like (2.16).

It can be proved that with limits $t = \infty$ either integral in (2.16) is equal to $\pi/2$. Otherwise Si(t) is not available in closed form (i.e. in terms of known functions of t), but it is, however, a tabulated function.

2.7 FURTHER FOURIER TRANSFORM PAIRS

At this stage we include just two more results, to end Table 2.2. (Other results can be derived as needed from the defining integral and properties.) Pair FT9 is the subject of Problem 2.3 at the end of the chapter, but we will discuss the implications and application of FT10.

Worked Example 2.7

Show that

$$\mathcal{F}\left\{\sum_{m=-\infty}^{\infty} \delta(t - mT_s)\right\} = f_s \sum_{m=-\infty}^{\infty} \delta(f - mf_s)$$

in which

$$f_s = \frac{1}{T_s}$$

Solution

We begin with some discussion. The function to be transformed is a train of unit impulses at points $t = mT_s$ on $-\infty < t < \infty$, equally spaced if T_s is a constant. It can therefore be regarded as a periodic impulse function, and we can regard T_s as a sampling interval inasmuch as multiplying a function $x(t)$ by this sum is equivalent to generating a sequence of sampled values of $x(t)$ at discrete points $t = mT_s$, illustrated in Fig. 2.7.

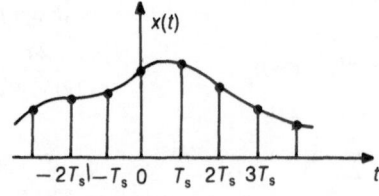

Fig. 2.7 Ideal sampling of a continuous signal.

In this sense, $\sum_{-\infty}^{\infty} \delta(t - mT_s)$ can be regarded as an ideal (because *instant*) sampling function. (In practice, sampling is not instant, however small the time taken to obtain a sampled value.)

As defined as an integral, the Fourier transform of a periodic transform does not exist, but we can proceed via the Fourier exponential *series* representation of the impulse function, and then transform that. As will be seen, the result is an impulse train in the *frequency* domain – which is more useful in the context of sampling than would be a sum of exponential functions (as implied by pair FT4 ᶠ Table 2.2).

More specifically, we can regard $\sum_{m=-\infty}^{\infty} \delta(t - mT_s)$ as the function $\delta(t)$ defined over one period, say $[-T_s/2, T_s/2]$ plus its periodic extensions. From Section 1.2 it follows that the coefficients X_n in the series

$$\sum_{n=-\infty}^{\infty} X_n e^{j2n\pi t/T_s}$$

are given by

$$X_n = \frac{1}{T_s} \int_{-T_s/2}^{T_s/2} \delta(t') e^{-j2\pi n t'/T_s} dt'$$

From the sifting property of the δ-function

$$X_n = \frac{1}{T_s} \cdot 1 = f_s$$

having 'picked out' the value of the exponential factor in the integrand at $t = 0$.

(The same result would be obtained by integrating over *any* interval of length T_s including an impulse $\delta(t - mT_s)$, as can be verified remembering that $e^{j2k\pi} = 1$ for any positive or negative integer k.)

As f_s is a constant, all coefficients X_n are equal, and we have the Fourier exponential series representation

$$\sum_{m=-\infty}^{\infty} \delta(t - mT_s) \sim f_s \sum_{n=-\infty}^{\infty} e^{j2n\pi t/T_s}$$

If we now transform the right-hand side of this expression using pair FT5 in which f_0 is replaced by $n/T_s = nf_s$, we obtain

$$\mathscr{F}\left\{ f_s \sum_{n=-\infty}^{\infty} e^{j2n\pi t/T_s} \right\} = f_s \sum_{n=-\infty}^{\infty} \delta(f - nf_s).$$

It is immaterial what notation is used for the summation integer, and replacing n by m in the last expression provides the required result.   ●

In this example we anticipated the replacement of a continuous signal $x(t)$ by a sequence of sampled values. Equally, we shall need to consider working with discrete values of transforms, which again implies a sampling process. This is why we have sought to identify an impulse train in the frequency domain.

2.8 GRAPHICAL ASPECTS OF CONVOLUTION

The evaluation of a convolution integral (in either the time or transform domain) may be relatively straightforward if the integral exists and the integrand factors are defined continuously and in terms of relatively simple functions.

Worked Example 2.8

Find $x_1 * x_2$ given $x_1(t) = e^{-bt} H(t-1)$ and $x_2(t) = e^{-a|t|}$, in which $0 < a < b$.

Solution

If we substitute into the convolution integral in the form

$$x(t) = x_1 * x_2 = \int_{-\infty}^{\infty} x_1(t') x_2(t - t') \, dt'$$

and note that $H(t' - 1) = 0$ if $t' < 1$, then

$$x_1 * x_2 = \int_{1}^{\infty} e^{-bt'} e^{-a|t - t'|} \, dt'$$

Now although the factor $e^{-a|t - t'|}$ is symmetrical (in that it is the same as $e^{-a|t' - t|}$), we have to stop and consider the fact that in this integration process t is a *parameter*, meaning a variable whose value is unspecified, and so the sign of $(t - t')$ has to be considered carefully before we decide how to interpret $|t - t'|$.

This is the first indication that graphical illustration of the integrand is important. The function $e^{-a|t|}$ has a maximum at $t = 0$. The function $e^{-a|t - t'|}$ has a maximum at $(t - t') = 0$, i.e. at $t' = t$. The

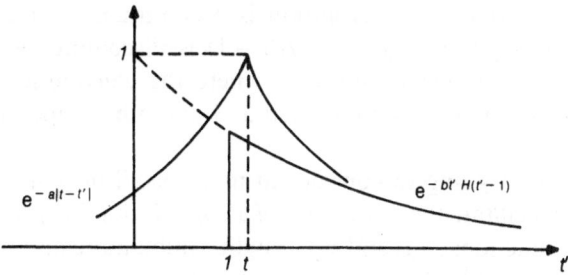

Fig. 2.8 The factors in the convolution integrand.

integrand factors, as functions of t', are sketched in Fig. 2.8. Here we have illustrated the case $t > 1$. The integrand is the product of the two factor values at any point t', and clearly if $t > 1$ this will involve *two* expressions for $e^{-a|t-t'|}$, one describing where the function is increasing (between 1 and t), the other equation describing the decreasing function (when $t' > t$). It follows that if $t > 1$ we should substitute

$$e^{-a|t-t'|} = e^{-a(t-t')} \qquad 1 < t' < t$$

and

$$e^{-a|t-t'|} = e^{+a(t-t')} \qquad t < t' < \infty$$

We therefore evaluate

$$\int_1^t e^{-bt'} e^{-a(t-t')} dt' + \int_t^\infty e^{-bt'} e^{+a(t-t')} dt'$$

Remembering that $0 < a < b$, it can easily be verified that

$$x(t) = x_1 * x_2 = \frac{e^{-at}}{b-a}[e^{-(b-a)} - e^{-(b-a)t}] + \frac{e^{at}}{b+a}[e^{-(b+a)t}] \qquad (2.17)$$

If the parameter $t < 1$, the point-by-point product involves only the decreasing branch of $e^{-a|t-t'|}$, and since $t < 1 < t'$ we substitute $e^{+a(t-t')}$ again, and integrate to obtain

$$x_1 * x_2 = \int_1^\infty e^{-bt'} e^{+a(t-t')} dt' = \frac{e^{at}}{b+a}[e^{-(b+a)}] \qquad (2.18)$$

If we consider the values of $x(t) = x_1 * x_2$ at the point $t = 1$, (2.17) and (2.18) both give

$$x(1) = \frac{e^{-b}}{b+a}$$

which means that the convolution is a continuous function, even though the function $x_1(t) = e^{-bt}H(t-1)$ is discontinuous at $t=1$. This can be expected in any case where the convolution integral exists; convolution can be regarded as a smoothing operation. ●

We are next going to consider in more detail how sketches can be used to facilitate the evaluation of a convolution integral, in cases where the functions $x_1(t)$ and $x_2(t)$ might exhibit more discontinuities than in the example above, and/or be defined as non-zero on intervals of finite length. This will involve intermediate sketches of (for example) $x_2(-t')$ and $x_2(t-t')$, if we adopt the same form of the convolution integral as above, as well as 'combined' sketches of $x_1(t')$ and $x_2(t-t')$.

We will use some of the recently defined functions to illustrate the process.

Worked Example 2.9

Sketch the functions $x_1(t') = \Pi(t'/4)\,\mathrm{sgn}\,t'$ and $x_2(t') = 2\Pi(t' - \tfrac{1}{2})t'$ and find the convolution

$$x(t) = x_1 * x_2 = \int_{-\infty}^{\infty} x_1(t')x_2(t-t')\,\mathrm{d}t'$$

Solution

The functions are sketched separately in Fig. 2.9. The first step towards constructing a sketch of $x_2(t-t')$ is called *folding* or *reversal*,

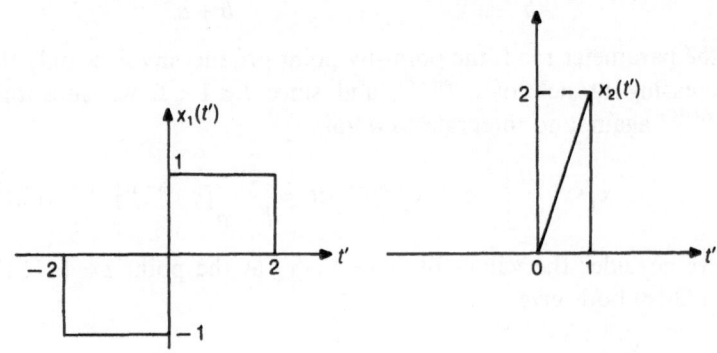

Fig. 2.9 Graphs of the functions to be convolved (Worked Example 2.9).

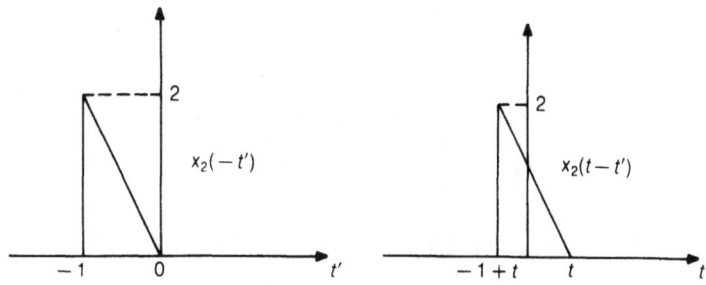

Fig. 2.10 Folding and shifting $x_2(t')$ to obtain the factor $x_2(t - t')$.

and is effected by changing the sign of t', giving $x_2(-t')$, whose graph is in Fig. 2.10. This is then shifted through a distance t on the t'-axis if we complete the construction of the second factor of the convolution integrand and sketch $x_2(t - t')$. (In Fig. 2.10 this is shown as if $t > 0$, but we have to consider all parameter values $-\infty < t < \infty$.)

Now consider the position of $x_2(t - t')$ in relation to that of $x_1(t')$. Both signals are non-zero on a finite range of values of t'. At points t' where either is (or both are) zero, their product is zero and there is no contribution to the convolution $x(t)$. Next, therefore, we illustrate the various possibilities which arise depending on different values of t.

First, if $t < -2$ there are no values of t' on $(-\infty, \infty)$ for which $x_1(t')x_2(t - t')$ is non-zero, as Fig. 2.11 shows, hence $x_1 * x_2 = 0$.

Similarly, if $x_2(t - t')$ is entirely 'to the right' of $x_1(t')$, as shown in Fig. 2.12, all point-by-point products are zero and $x_1 * x_2 = 0$. We see that this happens if $-1 + t > 2$, i.e. $t > 3$.

It is necessary, therefore, to consider only the evaluation of $x(t) \neq 0$ on $-2 < t < 3$ and the intervals on the t'-range where both $x_1(t')$

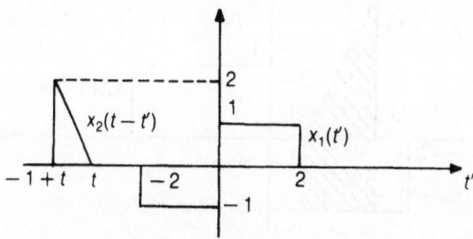

Fig. 2.11 The convolution is zero if $t < -2$.

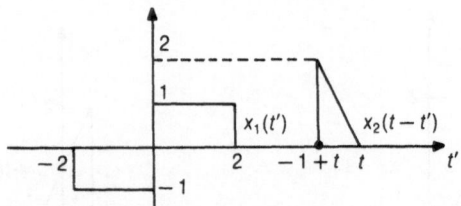

Fig. 2.12 All products are zero if $t > 3$.

and $x_2(t - t')$ are non-zero. Figures 2.13–2.17 show the effects of increasing t over the range of values $(-2, 3)$.

We have the type of 'overlap' shown in Fig. 2.13 if $-1 + t < -2$ and $t > -2$ i.e. $-2 < t < -1$. Noting that the equation of the second function is

$$x_2(t - t') = -2(t' - t)$$

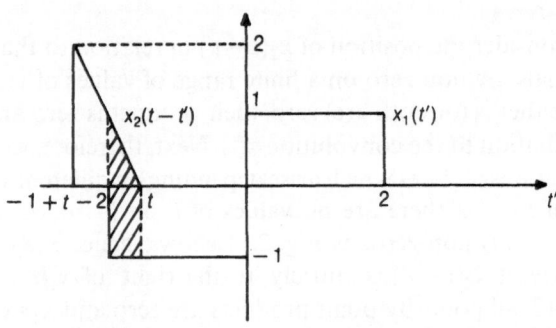

Fig. 2.13 Increasing t between -2 and -1.

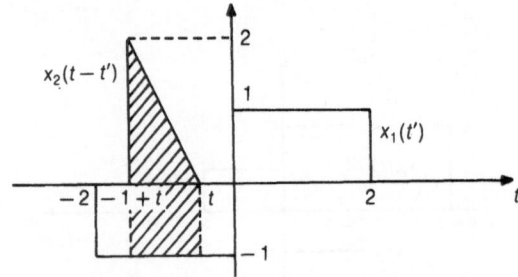

Fig. 2.14 Increasing t between -1 and 0.

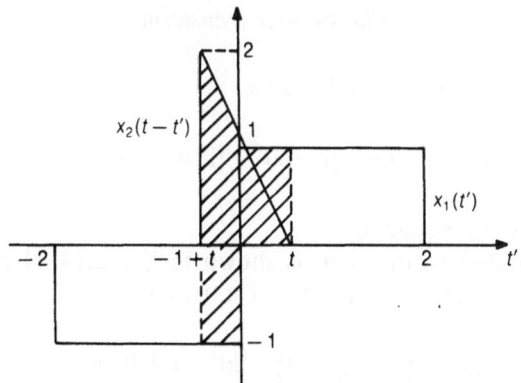

Fig. 2.15 Increasing t between 0 and 1.

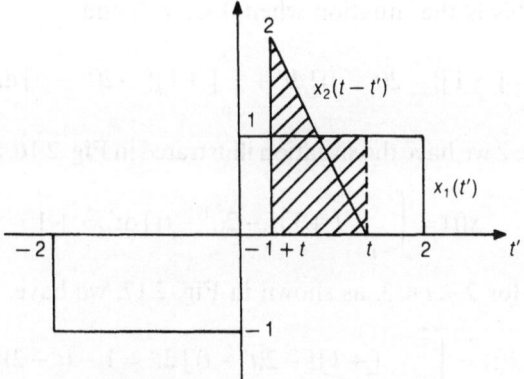

Fig. 2.16 Increasing t between 1 and 2.

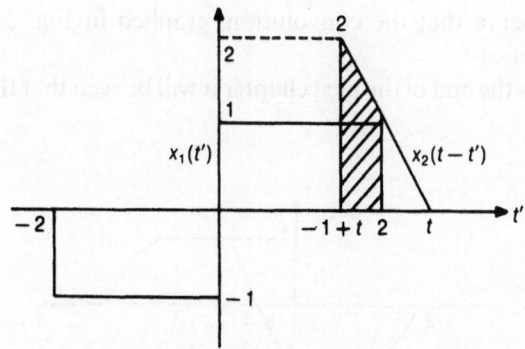

Fig. 2.17 Increasing t between 2 and 3.

the convolution integral reduces to

$$x(t) = \int_{-2}^{t} [-1][-2(t' - t)] \, dt' = -(t + 2)^2$$

(as can easily be verified).

A further shift to the right is shown in Fig. 2.14. Here we have $-1 + t > -2$ and $t < 0$, i.e. $-1 < t < 0$, and

$$x(t) = \int_{-1+t}^{t} [-1][-2(t' - t)] \, dt' = -1$$

Shifting further, as shown in Fig. 2.15, it is necessary to split the integral as the range of integration includes a change in definition of $x_1(t')$. This is the situation when $0 < t < 1$, and

$$x(t) = \int_{-1+t}^{0} [-1][-2(t' - t)] \, dt' + \int_{0}^{t} [+1][-2(t' - t)] \, dt' = 2t^2 - 1.$$

If $1 < t < 2$ we have the situation illustrated in Fig. 2.16, and obtain

$$x(t) = \int_{-1+t}^{t} [+1][-2(t' - t)] \, dt' = +1$$

Finally, for $2 < t < 3$, as shown in Fig. 2.17, we have

$$x(t) = \int_{-1+t}^{2} [+1][-2(t' - t)] \, dt' = 1 - (t - 2)^2$$

Summing up, five different expressions have been obtained for the convolution covering the range $-2 < t < 3$, outside which $x(t) = 0$. Again, observe that the convolution, graphed in Fig. 2.18, has no discontinuities.

Towards the end of the next chapter it will be seen that the concepts

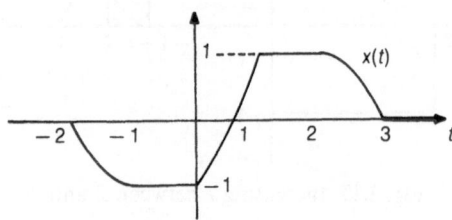

Fig. 2.18 The convolution $x(t) = x_1 * x_2$.

of folding and shifting are equally of use when it is wished to convolve two discrete sequences (of sampled values of two signals). ●

SUMMARY

In this chapter we have established various properties of the Fourier transform and also derived a selection of transform pairs, including the transforms of a number of commonly occurring functions, some of which are discontinuous. The significance of the duality property of the Fourier transform has been emphasized through application, and the transform of an impulse train has been discussed in anticipation of sampling a continuous function. Finally, we have shown how graphical detail helps interpret and evaluate a convolution integral.

PROBLEMS

2.1 From the defining integral of the Fourier transform, (2.1), prove the scaling property

$$\mathscr{F}\{x(at)\} = \frac{1}{|a|} X\left(\frac{f}{a}\right)$$

2.2 Use the translation property F5 of Table 2.1 to deduce that if

$$\mathscr{F}\{x(t)\} = X(f)$$

then

$$\mathscr{F}\{\cos(2\pi f_0 t)x(t)\} = \tfrac{1}{2}[X(f - f_0) + X(f + f_0)]$$

By considering, for example, $\delta(f - f_0) * X(f)$, show that the transform

$$\tfrac{1}{2}[X(f - f_0) + X(f + f_0)]$$

can be written in terms of convolution integrals in the domain of f. Use property F8 of Table 2.1 to deduce that

$$\mathscr{F}\{\cos 2\pi f_0 t\} = \tfrac{1}{2}[\delta(f - f_0) + \delta(f + f_0)]$$

(From which transform pair can the last result be derived directly?)

2.3 From the defining integral, (2.1), show that

$$\mathscr{F}\{e^{-a|t|}\} = \frac{2a}{(2\pi f)^2 + a^2}$$

in which $a > 0$.

2.4 Use either the inversion integral (2.2) or the duality and differentiation properties of Table 2.1 to show that

$$\mathscr{F}\{tx(t)\} = -\frac{1}{2\pi j}\frac{dX}{df}$$

2.5 Show that

$$\mathscr{F}\{\Pi(2t)\cos 2\pi t\} = \frac{1}{4}\left[\operatorname{sinc}\left(\frac{f-1}{2}\right) + \operatorname{sinc}\left(\frac{f+1}{2}\right)\right]$$

by using appropriate properties of the Fourier transform. Show that direct integration using (2.1) gives the result in the form

$$\mathscr{F}\{\Pi(2t)\cos 2\pi t\} = \frac{1}{\pi}\frac{\cos(\pi f/2)}{1 - f^2}$$

(Note that $e^{\pm j\pi/2} = \pm j$). Use the definition of the sinc function, (2.15), to verify that these results are the same.

2.6 If $x_1(t) = x_2(t) = \Pi(t/B)$, show that the convolution $x(t) = x_1*x_2 = B\Lambda(t/B)$. Verify from tables that this is consistent with the property

$$\mathscr{F}\{x_1*x_2\} = X_1(f)X_2(f).$$

2.7 Obtain the convolution of

$$x_1(t) = \Pi\left(\frac{t-1}{2}\right) \quad \text{and} \quad x_2(t) = H(t-4)$$

and show that your result can be expressed

$$x_1*x_2 = H(t-4)(t-4) - H(t-6)(t-6).$$

2.8 Using either version of the convolution integral, with

$$x_1(t) = 2e^{-4t}H(t) \quad \text{and} \quad x_2 = H(t-2)$$

obtain the result

$$x(t) = \begin{cases} x_1*x_2 = 0 & t < 2 \\ \frac{1}{2}[1 - e^{-4(t-2)}] & t > 2. \end{cases}$$

2.9 A circuit is described (with the usual notation) by the equation

$$\frac{dv}{dt} + \frac{1}{RC}v = \frac{i(t)}{C}$$

Use the differentiation property of the Fourier transform and Table 2.2 to show that if the input current $i(t) = \delta(t)$ then the ensuing voltage is given by

$$v(t) = \frac{1}{C}e^{-t/RC}H(t)$$

2.10 The Fourier cosine transform $\mathscr{F}_c\{x(t)\} = X_c(w)$ and the Fourier sine transform $\mathscr{F}_s\{x(t)\} = X_s(w)$ of a function $x(t)$ were defined in Section 1.6 by (1.15) and (1.17), respectively. Use these expressions and integrate by parts to show that

(i)
$$\mathscr{F}_c\left\{\frac{dx}{dt}\right\} = x(0) + wX_s(w)$$

(ii)
$$\mathscr{F}_c\left\{\frac{d^2x}{dt^2}\right\} = x'(0) - w^2X_c(w)$$

(iii)
$$\mathscr{F}_s\left\{\frac{dx}{dt}\right\} = -wX_c(w)$$

(iv)
$$\mathscr{F}_s\left\{\frac{d^2x}{dt^2}\right\} = wx(0) - w^2X_s(w)$$

2.11 Show that the Fourier sine and cosine transforms have the following properties:

(i)
$$\mathscr{F}_s\{tx(t)\} = -\frac{dX_c}{dw}$$

(ii)
$$\mathscr{F}_c\{tx(t)\} = \frac{dX_s}{dw}$$

2.12 (Modulation by a sinusoid)
Show that
$$\mathscr{F}_s\{x(t)\cos kt\} = [X_s(w+k) + X_s(w-k)]/2.$$

3

Discrete signals and transforms. The Z-transform and discrete convolution

Whether signals and spectra are initially in discrete or continuous form, to apply digital computational techniques we need discrete versions of both. The analogue–digital conversion process induces errors (and other errors occur in subsequent computation, such as involve numerical integration, and the truncation of infinite series).

In this chapter we shall first consider some of these matters, and the effects of sampling, and then proceed to introduce the Z-transform, which in effect is a discrete version of the Laplace transform.

No prior knowledge of the Z-transform will be assumed, although some readers might have encountered it (possibly defined as a series rather than as a derivation from the Laplace transform), and might have seen some applications – as in the description of system transfer functions. (Another application of the Z-transform is to be found in the analysis of the stability of a system, which depends on the location of the singularities of the transform in the complex plane of z. This is outside our present remit, and will not be pursued.) The Z-transform is ideally suited when handling sampled values (of inputs to and outputs from a linear system, for example) and provides a powerful method of solving linear difference equations. That is the subject of Chapter 4, and so here we just observe that linear differential equations must be replaced by equivalent discrete equations when the continuous-time variable t is replaced by a set of sampling times, $\{nT_s\}$. The chapter is concluded with a description of discrete convolution. It has been seen previously that multiplication in the frequency domain is associated with convolution in the time domain, and this can assist with both the analysis and synthesis of a transfer function.

The convolution sum (which is the equivalent of the continuous system convolution integral) is obtained using the Z-transform. (However, the discrete convolution expressions for other transforms have a similar structure.)

3.1 SAMPLING, QUANTIZATION AND ENCODING

That *sampling* an analogue signal at over-long intervals (too low a frequency) can lead to distortion should be self-evident. At the simplest level, interpolation between sampled values must not give a misleading idea of the form of the original (continuous-time) signal, as Fig. 3.1 suggests might happen. (Throughout, we shall concern ourselves only with a *constant* sampling interval, T_s, and a constant sampling frequency $f_s = 1/T_s$). In Section 3.5 we shall consider the question of maximum T_s (and hence minimum f_s) if distortion is to be avoided – as far as is possible in practice.

Sampling in the time domain has the effect of replacing the continuous variable t by the discrete variable nT_s, but amplitude (for example) still takes values $|x(nT_s)|$ on a continuous range. This also must be replaced by a discrete variable, and the process by which this is done is called *quantization*, whereby the value of a quantity defined continuously is assigned to a proximate number on a 'discrete' range of values. Quantization is illustrated in Fig. 3.2. Errors here can be regarded as rounding errors, and clearly will depend on the interval between the successive numbers available on the discrete range.

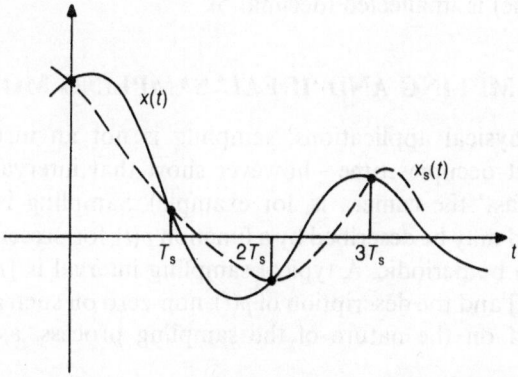

Fig. 3.1 Sampling at a rate which provides inadequate information.

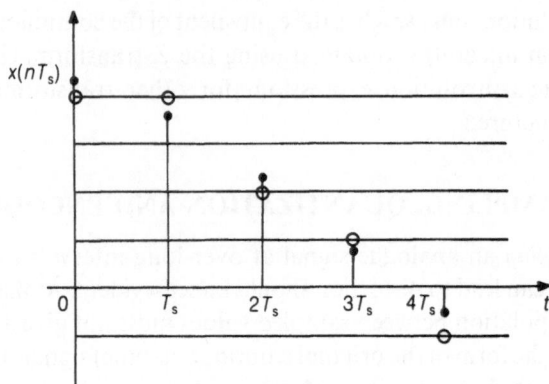

Fig. 3.2 Converting amplitude to discrete form.

Having converted both time and amplitude variables to discrete form, the final process in analogue–digital conversion is *encoding*. As this involves only the conversion of numbers to machine-acceptable form (i.e. binary), no further error is incurred. The conversion of a decimal number to bit form is achieved by expressing the former as the sum of successive powers of 2. Each successive power will be included in the sum once or not at all. For example, $13 = 8 + 4 + 1 = 2^3 + 2^2 + 0.2^1 + 2^0$, and so 13 in binary form is 1101. We might also mention *bit-reversal* at this point, an operation required in Chapter 7. This means writing the digits of a binary number in reverse order, with the result that, unless the binary number is palindromic, a different number is obtained. For example, if we reverse 1101 we obtain 1011 which, in decimal, is $2^3 + 2^1 + 2^0 = 11$, whereas 101 (palindromic) is unaffected (decimal 5).

3.2 SAMPLING AND 'IDEAL' SAMPLING MODELS

In most physical applications, sampling is not an instantaneous process, but occupies time – however short that interval might be (however 'fast' the camera is, for example). Sampling is an on/off process, and may be described by a function $p(t)$, for present purposes assumed to be periodic. A typical sampling interval is $[mT_s - (\tau/2), mT_s + (\tau/2)]$ and the description of $p(t)$, non-zero on such an interval, will depend on the nature of the sampling process, as shown in Fig. 3.3.

A simple example is provided by a periodic pulse train of amplitude

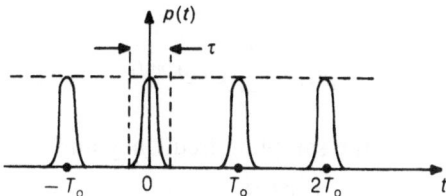

Fig. 3.3 Sampling action of duration τ.

1, which could be written

$$p(t) = \sum_{-\infty}^{\infty} \Pi\left(\frac{t - nT_s}{\tau}\right).$$

In general the equation

$$x_s(t) = p(t)x(t) \tag{3.1}$$

describes the set of sampled values, $\{x(nT_s)\}$. This is a product, and so any form of sampling in the time domain has an effect in the transform domain, which will be considered in later sections.

An 'ideal' sampling model would be one which is instantaneous $(\tau \to 0)$. The δ-function was first introduced in Table 1.1, where it was defined as a limit. The sifting property of the δ-function was given, and that could be taken as an alternative definition. (It should be emphasized that $\delta(t)$ is not a function in the conventional sense – it cannot be evaluated over a finite range of t-values, for example). In Section 2.7 we have already considered using a train of impulses as a sampling function (and also, using a Fourier series representation, obtained an expression for its Fourier transform). From (3.1), 'ideally sampled' values are given by

$$x_s(t) = \sum_{-\infty}^{\infty} \delta(t - nT_s)x(t) \tag{3.2}$$

3.3 THE FOURIER TRANSFORM OF A SAMPLED FUNCTION

As defined by an integral (2.6), strictly speaking the transform of a periodic function does not exist, but we can bypass that difficulty, as has been done previously.

Let us first represent the sampling function by a Fourier exponential series

$$p(t) \sim \sum_{-\infty}^{\infty} C_n e^{j2n\pi f_s t}$$

in which $f_s = 1/T_s$ is the sampling frequency and

$$C_n = \frac{1}{T_s} \int_{-T_s/2}^{T_s/2} p(t) e^{-j2n\pi f_s t} dt$$

From (3.1) the expression for sampled values can be written

$$x_s(t) = \sum_{-\infty}^{\infty} C_n x(t) e^{j2n\pi f_s t}$$

The spectrum of the sampled signal is the Fourier transform of this, namely

$$X_s(f) = \int_{-\infty}^{\infty} \left\{ \sum_{-\infty}^{\infty} C_n x(t) e^{j2n\pi f_s t} \right\} e^{-j2\pi f t} dt$$

Noting the presence of the exponential factor $e^{j2n\pi f_s t}$, we may invoke the translation property F5 of Table 2.1, replacing f_0 by nf_s. Then, if the (continuous) signal $x(t)$ has a transform $X(f)$, $x(t)e^{j2n\pi f_s t}$ has a transform $X(f - nf_s)$ and the previous equation can be written

$$X_s(f) = \sum_{-\infty}^{\infty} C_n X(f - nf_s) \tag{3.3}$$

Note that $X_s(f)$ is a function of the *continuous* variable f, and the suffix s indicates that we are referring to the transform of a time-domain sampled signal. Not until Chapter 5 shall we consider a discrete form of the Fourier transform in any detail. (That will involve replacing f by a discrete variable.)

In (3.3) the terms $X(f - nf_s)$ are repeats of $X(f)$, shifted through multiples of the sampling frequency f_s on the f-axis. As C_n is in general a function of n, they are differently weighted repeats and therefore $X_s(f)$ is not periodic. The effect in the transform domain of sampling in the time domain is illustrated in Figs. 3.4 and 3.5. The position as shown has been simplified in two respects. First, $X(f)$ in general is a complex quantity and so really we have sketched $|X(f)|$. Second, but more significantly, we have suggested that $X(f)$ is zero outside a finite range $(-f_h, f_h)$. That implies that $x(t)$ is a *band-limited* signal, meaning that the highest frequency component present is associated with some $f = f_h$. This is normally an approxi-

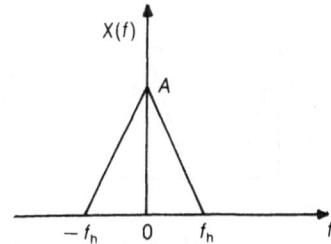

Fig. 3.4 The transform of $x(t)$.

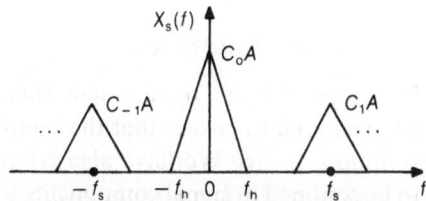

Fig. 3.5 The transform of $x_s(t)$.

mation to the true situation, and in practice we are regarding as negligible the energy content of the signal components for which $|f| > f_h$.

3.4 THE SPECTRUM OF AN 'IDEALLY SAMPLED' FUNCTION

In Sections 2.7 and 3.2 we considered sampling by means of a train of unit impulses. (Use of the adjective 'unit' reflects the fact that the impulse is the limit of a function which satisfies the equation $\int_{-\infty}^{\infty} f(t)\mathrm{d}t = 1, f(t) \geqslant 0$.) It was shown in Section 2.7 that all coefficients C_n in the Fourier exponential series representation of the sampling function are in this case equal, and equal to the sampling frequency. Equation (3.3) therefore takes the special form

$$X_s(f) = f_s \sum_{-\infty}^{\infty} X(f - nf_s) \qquad (3.4)$$

and the shifted repeats of $X(f)$ are now equally weighted. Thus, the effect of ideal sampling in the time domain is to introduce periodicity in the frequency domain, associated with an amplification factor f_s, as is shown in Fig. 3.6.

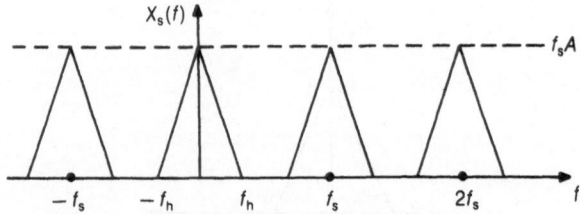

Fig. 3.6 The periodic transform of an ideally sampled function, $x_s(t)$.

3.5 ALIASING

We have already recognized the requirement that the sampling frequency f_s be large enough to ensure that the sampled values give a reasonable description of $x(t)$. We have also referred to the fact that attention is to be confined to signal components with frequencies in the range $|f| < f_h$. The two are associated, and we can illustrate what happens should the selected f_s be too small. Should $f_s - f_h$ be less than f_h, then the graphs of the terms $f_s X(f - nf_s)$ are not distinctly spaced on the f-axis, but 'overlap' – as we have shown in Fig. 3.7 with $n = 0$ and $n = 1$. This happens if $f_s < 2f_h$. The effect is called *aliasing* and $2f_h$ is called the *Nyquist rate* (of sampling).

The perceived spectrum $X_s(f)$ is distorted because of addition of terms in overlap intervals, as illustrated in Fig. 3.8. The form of the original transform $X(f)$ is not identifiable, and whatever section of $X_s(f)$ is examined, inversion will not lead to the correct recovery of the time-domain function.

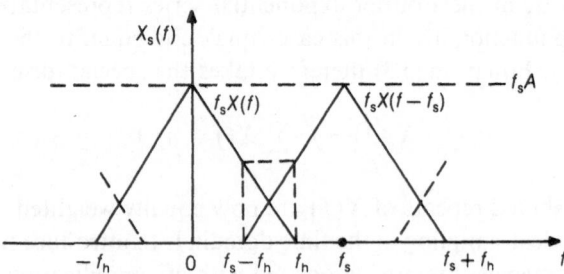

Fig. 3.7 Aliasing: interference between terms in $X_s(f)$.

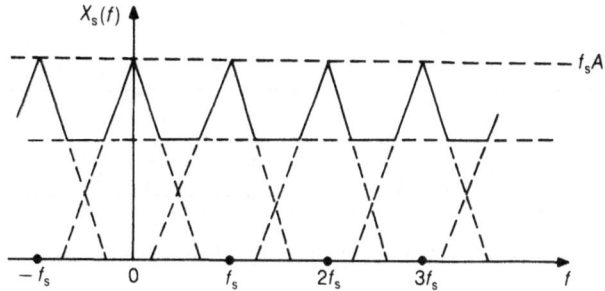

Fig. 3.8 The resultant transform $X_s(f)$ showing aliasing.

Worked Example 3.1

The spectrum, $\mathscr{F}\{x(t)\} = X(f)$ is as shown in Fig. 3.9, in which $f_h = 20$. Illustrate the effects in the frequency domain of instant sampling in the time domain at rates $f_s = 30, 40$ and 50.

Solution

The terms in $X_s(f)$ are each of the form $f_s X(f - nf_s)$, non-zero over the interval $[nf_s - f_h, nf_s + f_h] = [nf_s - 20, nf_s + 20]$.

If $f_s = 30 < 2f_h = 40$, there is aliasing, as shown in Fig. 3.10. Summation over intervals of 'overlap', such as $[10, 20]$ on the axis where two terms are non-zero, results in a completely inappropriate spectrum. The resultant spectrum is shown in Fig. 3.11.

If $f_s = 40 = 2f_h$, we are sampling at the Nyquist rate, the critical frequency, and the spectrum is as illustrated in Fig. 3.12.

If $f_s = 50 > 2f_h = 40$, then the weighted repeats of $X(f)$ are separated by intervals where $X_s(f)$ is zero. There is no 'distortion'. The maximum amplitude is scaled by a factor 50. We leave it to the

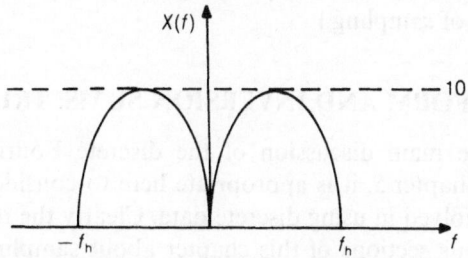

Fig. 3.9 A hypothetical spectrum of a band-limited function, $x(t)$.

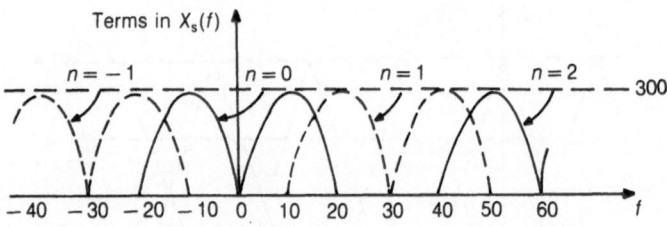

Fig. 3.10 Terms $f_s X(f - nf_s)$ when f_s is below the Nyquist rate.

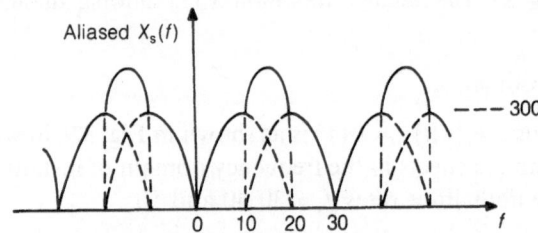

Fig. 3.11 The 'perceived' spectrum $X_s(f)$.

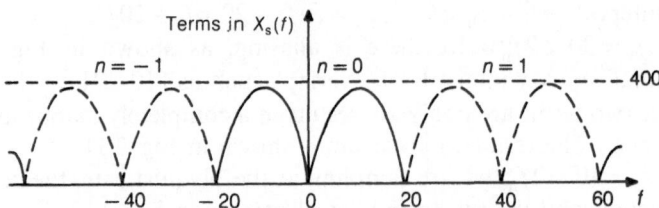

Fig. 3.12 Minimum frequency precluding aliasing.

reader to sketch that situation. (Note that any 'inversion' process must compensate for the amplification factor f_s introduced as a consequence of sampling.) ●

3.6 TRANSFORM AND INVERSION SUMS: TRUNCATION

Although the main discussion of the discrete Fourier transform appears in Chapter 5, it is appropriate here to consider further the processes involved in using discrete data. Clearly the remarks made in the previous sections of this chapter about sampling a function $x(t)$ apply equally to any transform $X(f)$ (be this a Fourier transform

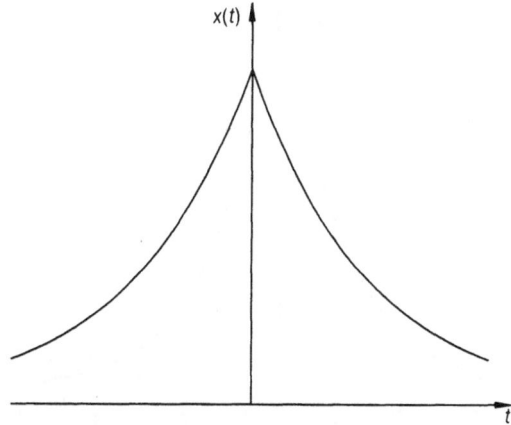

Fig. 3.13 A hypothetical signal, $x(t)$.

or not), which must also be sampled to use digital inversion techniques.

Also, any transform or inversion integral will be replaced by some form of series, depending on the method of numerical integration adopted and, as digital computation requires that the number of terms to be summed be finite, truncation of any infinite series is necessary. The number of sampled values of either $x(t)$ or $X(f)$ used has to be restricted.

Some of the remarks made in the rest of this section will be amplified in Section 3.7, particularly concerning truncation. We begin by considering an analogue signal $x(t)$ and its transform $X(f)$, making no assumption that either is band-limited, as Figs. 3.13 and 3.14 indicate. We have already obtained the transform of the unit impulse sampling function, now illustrated in Figs. 3.15 and 3.16.

Remembering that multiplication in the time domain is associated with convolution in the frequency domain, and conversely, in addition to the expression obtained for the transform of an ideally sampled signal in Section 3.4, we can write

$$X_s(f) = f_s \sum_{-\infty}^{\infty} X(f - nf_s) = X(f)*P(f)$$

The consequences of sampling are shown in Figs. 3.17 and 3.18, and the captions include the multiplications and convolutions associated with the process. Apart from the induced periodicity in $X_s(f)$, remember that aliasing is inevitable if a function is not band-limited.

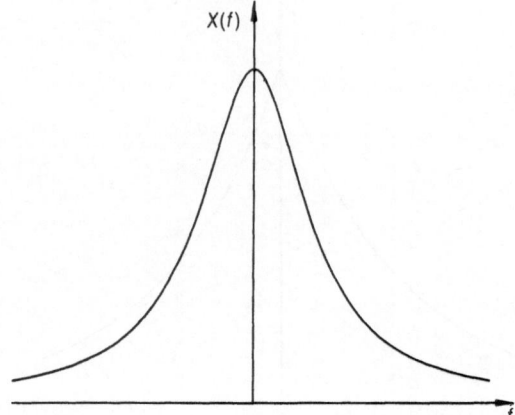

Fig. 3.14 A hypothetical transform, $X(f)$.

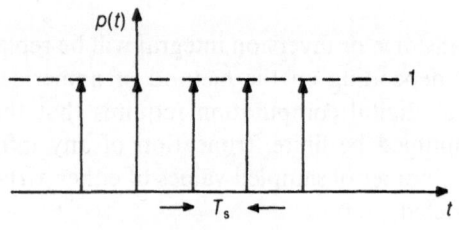

Fig. 3.15 $p(t) = \sum\limits_{-\infty}^{\infty} \delta(t - mT_s)$.

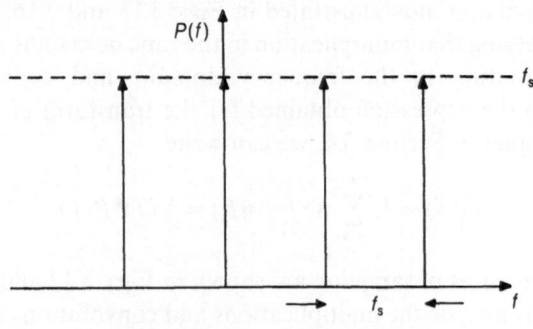

Fig. 3.16 $P(f) = f_s \sum\limits_{-\infty}^{\infty} \delta(f - nf_s)$.

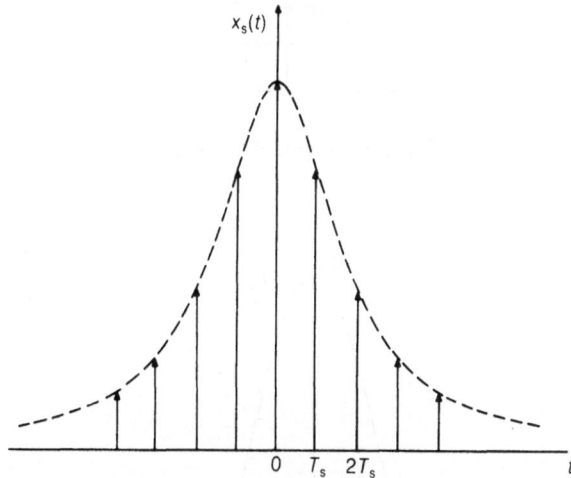

Fig. 3.17 The sampled signal $x_s(t) = p(t)x(t)$.

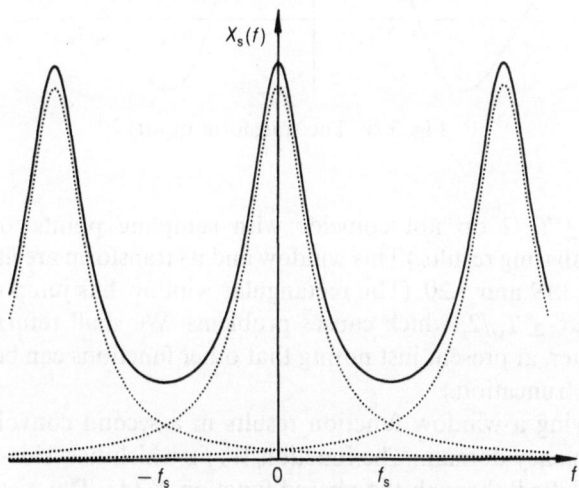

Fig. 3.18 The (aliased) spectrum $X_s(f) = X(f)*P(f)$.

We next come to the matter of truncating the sequence of sampled values $x_s(t)$. This can be achieved by the process called *windowing*, and involves multiplying $x_s(t)$ by a function $w(t)$ which is zero outside some finite range. A particularly simple function is the rectangular window $w(t) = \Pi(t/T_w)$, whose transform we know to be $W(f) = T_w \operatorname{sinc}(T_w f)$. (The window width, T_w, should be chosen so that

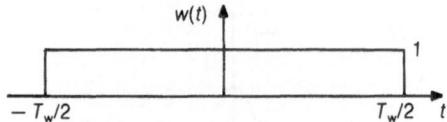

Fig. 3.19 A possible window.

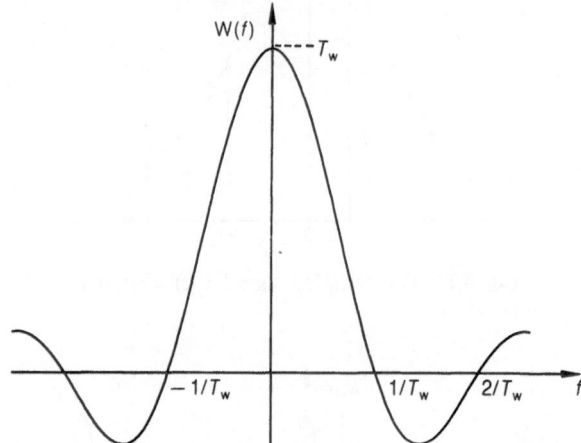

Fig. 3.20 The transform of $w(t)$.

points $\pm T_w/2$ do not coincide with sampling points, otherwise further aliasing results.) This window and its transform are illustrated in Figs. 3.19 and 3.20. (The rectangular window has jump discontinuities at $\pm T_w/2$, which causes problems. We shall return to this point later, at present just noting that other functions can be used to achieve truncation.)

Applying a window function results in a second convolution in the frequency domain. The function $W(f)$, which has 'side-lobes', is being 'shifted' through the aliased function $X_s(f)$. The picture then produced shows ripple effects, and error in the form of *spectral leakage*, meaning that some spectral energy is erroneously transferred between frequencies. Figures 3.21 and 3.22 show what happens.

The transform of the truncated sequence of sampled values of $x(t)$ is still a function of the continuous variable f, and so a similar process is required to invert it numerically. In computation, N time samples and N frequency samples are used. (To achieve this in the frequency domain we would sample at intervals of length $1/T_w$,

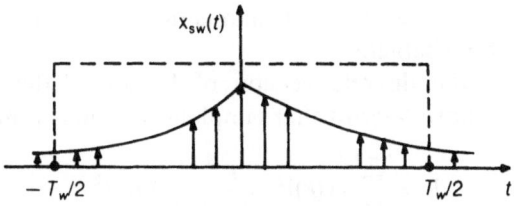

Fig. 3.21 The windowed sampled signal $x_{sw}(t) = w(t)p(t)x(t)$.

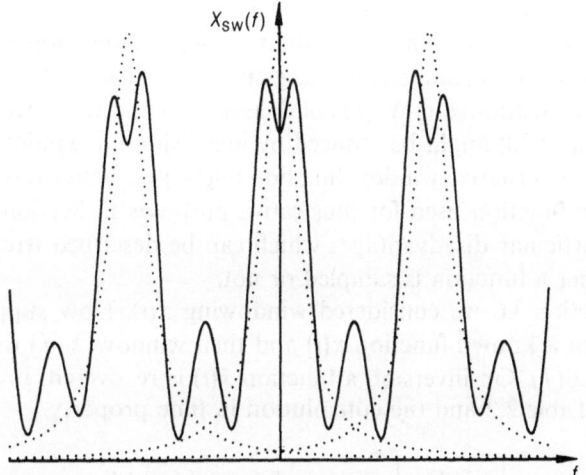

Fig. 3.22 The spectrum of $x_{sw}(t)$: $X_{sw}(f) = W(f)*P(f)*X(f)$.

where T_w is the width of the window applied in the time domain, and apply a window of length $f_s = 1/T_s$, where T_s is the interval between samples in the time domain.)

3.7 WINDOWING: BAND-LIMITED SIGNALS AND SIGNAL ENERGY

Normalized signal energy is defined by the equation

$$E = \int_{-\infty}^{\infty} |x(t)|^2 \, dt$$

and average power is half this. Energy may also be obtained from the spectrum, because from the definition above we can show that

$$E = \int_{-\infty}^{\infty} |X(f)|^2 \, df$$

This is called Parseval's theorem, and is the subject of Problem 3.3 at the end of the chapter.

Anticipating the discrete version of Parseval's theorem from Chapter 5, if we have N terms after sampling and windowing, then

$$E = \sum_{k=0}^{N-1} |x(k)|^2 = \frac{1}{N} \sum_{n=0}^{N-1} |X(n)|^2 \tag{3.5}$$

where $x(k)$ is the kth time sample and $X(n)$ is the nth element of the *discrete* transform.

In general, a signal is not band-limited, and (in either domain) truncation means that some signal energy is 'lost'. The sort of distortions and ripples exhibited in $X_{sw}(f)$ of the sampled and windowed signal $x_{sw}(t)$ (Fig. 3.20) might be reduced by increasing the window length T_w, or an alternative window function might give better results. The rectangle function used for illustrative purposes in Section 3.6 has some particular disadvantages which can be described irrespective of whether a function is sampled or not.

In Section 3.6 we considered windowing $x(t)$. Now suppose we transform a known function $x(t)$ and then window $X(f)$ to obtain $\Pi(f/f_s)X(f)$. On inversion a function $\tilde{x}(t)$ is recovered. From pair FT2 of Table 2.2 and the convolution in time property,

$$\tilde{x}(t) = \int_{-\infty}^{\infty} x(t-t')f_s \operatorname{sinc}(f_s t')dt'$$

This will be an approximation to $x(t)$. Thinking in terms of convolving graphically, shifting $x(t-t')$ through the sinc function and its side-lobes means that the graph of $\tilde{x}(t)$ will have oscillatory features. This is worsened in the region of any point t at which $x(t)$ has a discontinuity.

These effects are analogous to the oscillations found in the representation of a function by a truncated Fourier series, which can be particularly marked in the vicinity of a discontinuity (*Gibbs's phenomenon*).

Increasing window width ensures that the energy content of the windowed function is increased, but leakage is affected by both the width of the main lobe and the amplitudes of the side-lobes of a sinc function. Sinc functions appear in the Fourier transforms of other functions (see Problem 2.5, for instance) and by using a combination of sinc functions, it might be possible to reduce some of the problems associated with the rectangular window. Also, employing a function which has no such jump discontinuities should help. We include here just one example, in both its continuous and discrete forms.

The Hanning window

If applied in the frequency domain, this can be defined by the equation

$$W(f) = \Pi\left(\frac{f}{f_s}\right)\frac{1}{2}\left[1 + \cos\left(\frac{2\pi f}{f_s}\right)\right] \qquad (3.6)$$

which has inverse

$$w(t) = \frac{1}{2}\left[f_s\operatorname{sinc}(f_s t) + \frac{1}{2}\operatorname{sinc}\left(f_s\left[t + \frac{1}{f_s}\right]\right) + \frac{1}{2}\operatorname{sinc}\left(f_s\left[t - \frac{1}{f_s}\right]\right)\right]$$

The function $W(f)$ so defined has no discontinuity at $f = \pm f_s/2$, and the amplitudes of the side-lobes of $w(t)$ are less than those of the side-lobes of $f_s\operatorname{sinc}(f_s t)$.

The preceding can be rewritten for windowing in the time domain.

The discrete Hanning window, in a form suitable for application in the time domain, can be defined by the equation

$$w(k) = \frac{1}{2}\left[1 - \cos\left(\frac{2k+1}{N}\right)\right] \qquad (3.7)$$

for $k = 0, 1, \ldots, N - 1$. (It is unfortunately the case that references

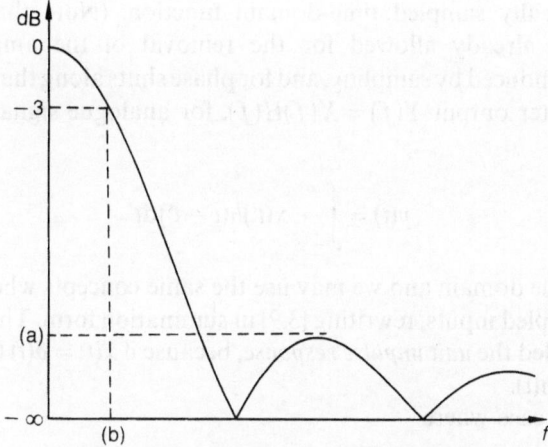

Fig. 3.23 Window amplitude in decibels.

and texts do not share a standard definition of window functions. What *is* common among the definitions of continuous and discrete Hanning windows is the presence of a factor $[1 \pm \cos(\text{something})]/2$. Most, but not all, references use the signs plus and minus as in (3.6) and (3.7).)

It should be pointed out that a great many window functions are available, and that they are usually compared (in relation to energy content) by looking at normalized amplitude (measured in decibels) as a function of frequency. Specifications usually give the amplitude of the first side-lobe, and give the frequency at which signal amplitude is 3 dB below the maximum. These are shown as (a) and (b), respectively, in Fig. 3.23. The 3 dB level gives (by implication) an idea of the number of sampled values, equivalent to an idea of the proportion of signal energy, included.

An ideal reconstruction filter

To put together a number of ideas and results developed in this and previous chapters we conclude this section with a discussion of an *ideal reconstruction filter*. An ideal low-pass filter rejects all input frequencies greater than some cut-off frequency. (In other words, *inter alia* it band-limits an input.) For example, consider adopting a transfer function

$$H(f) = T_s[\Pi(f/f_s)]e^{-j2\pi n T_s f} \qquad (3.8)$$

and applying this filter to invert the transform $X_s(f) = f_s \sum_{-\infty}^{\infty} X(f - nf_s)$ of an ideally sampled time-domain function. (Note that in (3.8) we have already allowed for the removal of the amplification factor f_s induced by sampling, and for phase shifts along the f-axis.)

The filter output $Y(f) = X(f)H(f)$, for analogue signals, corresponds to

$$y(t) = \int_{-\infty}^{\infty} x(t')h(t - t')dt' \qquad (3.9)$$

in the time domain and we may use the same concepts when dealing with sampled inputs, rewriting (3.9) in summation form. The function $h(t)$ is called the *unit impulse response*, because if $x(t) = \delta(t)$ the output is $y(t) = h(t)$.

In the case where

$$x_s(t) = \sum_{-\infty}^{\infty} \delta(t - nT_s)x(t) = \sum_{-\infty}^{\infty} \delta(t - nT_s)x(nT_s) \qquad (3.10)$$

the response to the impulse $x(0)\delta(t)$, with $H(f)$ as defined in (3.8), is given by putting $n = 0$ in the latter and quoting Fourier pair FT2 of Table 2.2. We obtain

$$x(0)h(t) = x(0)T_s f_s \operatorname{sinc}(f_s t) = x(0) \operatorname{sinc}(f_s t)$$

For $n \neq 0$ in (3.8) and (3.10), the delay property F4 of Table 2.1 gives the response to the nth sample as $x(nT_s)\operatorname{sinc}[f_s(t - nT_s)]$, and so the output from the reconstruction (inversion) filter is obtained in the form

$$y(t) = \sum_{-\infty}^{\infty} x(nT_s)\operatorname{sinc}[f_s(t - nT_s)] \qquad (3.11)$$

The output is continuous in time, and is a 'reconstruction' of $x(t)$. In general, accuracy depends on the adequacy of the time-domain sampling process and on the choice of transfer function $H(f)$. With this choice of filter, and sampling at a rate higher than the Nyquist rate, (3.11) is a statement of what is referred to as the *sampling theorem*, and we put $y(t) = x(t)$. The equation then states that the band-limited signal $x(t)$ is exactly specified in terms of its sampled values. Each term in the series is a sinc function whose maximum is at $t = nT_s$, weighted by the sample value $x(nT_s)$, as shown in Fig. 3.24.

In practice, the infinite limits of summation in (3.11) must be replaced by finite integers, processes are not 'ideal', and we put $y(t) \simeq x(t)$.

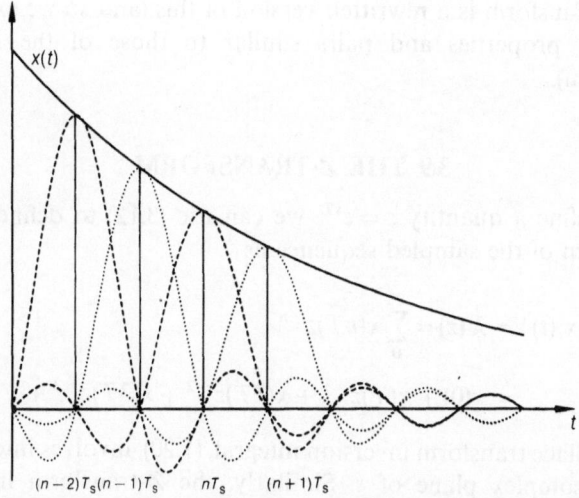

Fig. 3.24 Continuous signal $x(t)$ in terms of its sampled values.

3.8 THE LAPLACE TRANSFORM
OF A SAMPLED SIGNAL

For present purposes, suppose that $x(t) = 0$ on $t < 0$, that $x(t)$ is continuous at sampling points, and that ideally sampled values have been obtained so that

$$x_s(t) = \sum_{-\infty}^{\infty} x(t)\delta(t - T_s) = \sum_0^{\infty} x(nT_s)\delta(t - nT_s)$$

The Laplace transform of the sampled signal is

$$\mathcal{L}\{x_s(t)\} = \int_0^{\infty} x_s(t)e^{-st}\,dt = \sum_0^{\infty} x(nT_s) \int_0^{\infty} \delta(t - nT_s)e^{-st}\,dt$$

By the sifting property of the delta function, then

$$\mathcal{L}\{x_s(t)\} = \sum_0^{\infty} x(nT_s)e^{-snT_s}$$

To avoid any possible confusion concerning the use of s as the Laplace transform parameter, we will drop the use of s as a suffix (indicating sampling) on the right-hand side of this expression and write

$$\mathcal{L}\{x_s(t)\} = \sum_0^{\infty} x(nT)e^{-snT} \tag{3.12}$$

The Z-transform is a rewritten version of this (and so we expect to establish properties and pairs similar to those of the Laplace transform).

3.9 THE Z-TRANSFORM

If we define a quantity $z = e^{sT}$, we can use (3.12) to define the Z-transform of the sampled sequence as

$$\mathcal{Z}\{x_s(t)\} = X(z) = \sum_0^{\infty} x(nT)z^{-n}$$

$$= x(0) + x(T)z^{-1} + x(2T)z^{-2} + x(3T)z^{-3} + \cdots \tag{3.13}$$

The Laplace transform inversion integral, (1.20), involves integration in the complex plane of s. Similarly, the Z-transform inversion integral involves integration round a circle in the complex plane of z:

$$x(nT) = \frac{1}{2\pi j} \oint X(z)z^{n-1} dz \qquad (3.14)$$

This is included without proof for completeness only, because other methods of inversion will be used, as in the case of the Laplace transform, including the development of tables of transform pairs and properties.

Worked Example 3.2

Find the Z-transform of the following sequences:

(i) $x(0) = 1, x(nT) = 0$ if $n \neq 0$
(ii) $x(nT) = 1$
(iii) $x(nT) = e^{-anT}$
(iv) $x(nT) = nTe^{-anT}$

Solution

From (3.13),

(i) $X(z) = x(0) = 1$
(ii) $X(z) = 1 + z^{-1} + z^{-2} + z^{-3} + \cdots$
This series can be summed either by recognizing that it is the binomial series expansion of $(1 - z^{-1})^{-1}$, or that it is a geometric progression, $a + ar + ar^2 + ar^3 \ldots$, with first term $a = 1$, and common ratio $r = z^{-1}$. The sum to infinity of a converging geometric progression is $a/(1 - r)$. Either way, the Z-transform of a unit impulse train on $t \geqslant 0$ is $X(z) = 1/(1 - z^{-1})$.
(iii) $X(z) = 1 + e^{-aT}z^{-1} + e^{-2aT}z^{-2} + e^{-3aT}z^{-3} \cdots$
Using the same approach as in (ii),

$$X(z) = \frac{1}{1 - e^{-aT}z^{-1}}$$

(iv) We could approach this directly:

$$\mathscr{Z}\{nTe^{-anT}\} = 0 + Te^{-aT}z^{-1} + 2Te^{-2aT}z^{-2} + 3Te^{-3aT}z^{-3} + \cdots$$
$$= Te^{-aT}z^{-1}(1 + 2e^{-aT}z^{-1} + 3e^{-2aT}z^{-2} + \cdots)$$

Comparing the quantity in brackets with the binomial expansion of

$(1 - x)^{-2}$, then

$$X(z) = \frac{Te^{-aT}z^{-1}}{(1 - e^{-aT}z^{-1})^2}$$

Alternatively, having established in (iii) that the Z-transform of the sequence $\{e^{-anT}\}$ is $(1 - e^{-aT}z^{-1})^{-1}$, differentiation with respect to the parameter a in both the time and the transform domain will give the same result in the form

$$\mathscr{L}\{-nTe^{-anT}\} = \frac{-Te^{-aT}z^{-1}}{(1 - e^{-aT}z^{-1})^2}$$

Using such results and techniques, we will be able to establish further Z-transform pairs and properties.   ●

3.10 INPUT–OUTPUT SYSTEMS AND TRANSFER FUNCTIONS

The question of parameter identification arose in Worked Example 1.8, where from observed inputs and outputs it proved possible to characterize a linear system by a transfer function described in terms of the Laplace transform. (In that example we obtained a transfer *matrix*.)

Other transforms, including discrete transforms, are available. For example, with appropriate weighting attached to previous inputs, present input, and previous outputs (should feedback be an issue), we could write

$$y(nT) = \sum_{k=0}^{p} b_k x(nT - kT) - \sum_{k=1}^{q} a_k y(nT - kT) \qquad (3.15)$$

This is a linear difference equation, and is the discrete equivalent of a linear differential equation in $y(t)$, the response of a system to a known input $x(t)$. On the left, we have the present output. On the right we have present and previous inputs, and previous outputs. The unspecified upper summation limits, say integers p and q, will depend on the extent to which previous performance affects present performance (as do the weights $\{a_k\}$ and $\{b_k\}$), and will also reflect the need to deal with *finite* sums. Should the Z-transform be employed, there will be immediate recourse to sampled values, and we shall have a description of the transfer function from the equation $Y(z) = H(z)X(z)$. If this is to be obtained from (3.15), then the first requirement is to establish a shift or delay property for the Z-transform, in order to transform the terms $x(nT - kT)$ and $y(nT - kT)$ appearing in the right-hand side summations. Anticipating this, if (3.15) is transformed and all output terms are collected on the left of the equation, we will obtain the transfer function in the form

$$H(z) = \frac{\{b_0 + b_1 z^{-1} + \cdots + b_r r^{-p}\}}{\{1 + a_1 z^{-1} + a_2 z^{-1} + \cdots + a_m z^{-q}\}} = \frac{N(b_k, z^{-1})}{D(a_k, z^{-1})} \qquad (3.16)$$

in which the numerator, N, and denominator, D, are polynomials in z^{-1} of degree p and q, respectively. Justification of the delay property used in arriving at (3.16) is addressed immediately in the next section (see (3.19)).

3.11 PROPERTIES OF THE Z-TRANSFORM

The manner in which the delay properties of the Z-transform are derived from its definition as

$$\mathscr{Z}\{x(n)\} = \sum_{n=0}^{\infty} x(n)z^{-n} = X(z) \qquad (3.17)$$

is not dissimilar from the way in which the shift properties of the Laplace transform are derived from its defining integral. (Note that in (3.17) we have abbreviated notation further for present purposes by putting $x(nT) = x(n)$, to indicate sampled values.)

Multiplying (3.17) by z^{-m}, in which m is some fixed integer, gives

$$z^{-m}X(z) = \sum_{n=0}^{\infty} x(n)z^{-(n+m)}$$

Substituting $k = (n + m)$ then gives

$$z^{-m}X(z) = \sum_{k=m}^{\infty} x(k - m)z^{-k}$$

which can be written

$$z^{-m}X(z) = \sum_{k=0}^{\infty} x(k - m)z^{-k} - \sum_{k=0}^{m-1} x(k - m)z^{-k}$$

It does not matter what letter is used for the dummy integer in a sum. If in the last expression we put $k = n$ in the infinite sum and $(k - m) = -p$ in the finite sum we obtain

$$z^{-m}X(z) = \sum_{n=0}^{\infty} x(n - m)z^{-n} - \sum_{p=m}^{1} x(-p)z^{-m+p}$$

Looking at this and at (3.17), the first term on the right, by definition,
is the Z-transform of the sequence of sampled values shifted through
mT to the right on the time axis. The second sum is a finite sum,
the terms of which can be summed in reverse order and in which
we can replace p by k. At the same time, since m is fixed, we may
write z^{-m} as a factor before the summation sign. With these modifications, the last equation can be rearranged to read

$$\mathscr{Z}\{x(n-m)\} = z^{-m}\left[X(z) + \sum_{k=1}^{m} x(-k)z^k \right] \qquad (3.18)$$

This is a shift theorem for the Z-transform, in its most general form,
and appears as Z1(i) in Table 3.1. The presence of values $x(-k)$ might
seem inconsistent with the fact that at the beginning of Section 3.8
we considered the one-sided Laplace transform of a signal for which
$x(t) = 0$ on $t < 0$. However, when replacing differential equations by

Table 3.1 Z-transform properties, $X(z) = \mathscr{Z}\{x(n)\}$

Property	Time domain	Transform domain
Z1 Delay		
(i) forward	$x(nT - mT)$	$z^{-m}\left[X(z) + \sum_{k=1}^{m} x(-kT)z^k \right]$
(ii) backward	$x(nT + mT)$	$z^{m}\left[X(z) - \sum_{k=0}^{m-1} x(kT)z^{-k} \right]$
Z2 Damping		
$(a > 0)$ (i)	$x(nT)\,e^{-anT}$	$X(e^{aT}z)$
$(\alpha > 0)$ (ii)	$x(nT)\alpha^{-nT}$	$X(\alpha^{T}z)$
Z3 Initial value		
theorem	$x(0)$	$\lim\limits_{z \to \infty} X(z)*$
Final value		
theorem	$x(\infty)$	$\lim\limits_{z \to 1} (1 - z^{-1})X(z)*$
Z4 Convolution[†]		
	$x_1 * x_2$	$X_1(z)X_2(z)$
	$= \sum\limits_{m=0}^{n} x_1(mT)x_2(nT - mT)$	

* Results are conditional on the existence of these limits
† See Section 3.14.

difference equations (Chapter 4), if derivatives are approximated by expressions such as

$$\frac{dx}{dt} = \frac{x_n - x_{n-1}}{T}$$

rather than by equations of the form

$$\frac{dx}{dt} = \frac{x_{n+1} - x_n}{T}$$

then initial values such as $x(-1)$, $x(-2)$ might appear in the converted description of a system. For that reason, we provide for such values in (3.18) as it is really immaterial where, on the time axis, we start counting sampled values $x(nT)$, from some starting value of n. If that provision is not necessary, then (3.18) reduces to

$$\mathscr{Z}\{x(n-m)\} = z^{-m}X(z) \tag{3.19}$$

This is the property used in arriving at the expression for the transfer function $H(z)$ of (3.16) on putting $x(-k) = 0$ for all $k > 0$.

Otherwise, (3.18) gives expressions such as

$$\mathscr{Z}\{x(n-1)\} = z^{-1}X(z) + x(-1)$$

and

$$\mathscr{Z}\{x(n-2)\} = z^{-2}X(z) + z^{-1}x(-1) + x(-2)$$

The backwards equivalent of the property described by (3.18) is stated as follows:

$$\mathscr{Z}\{x(n+m)\} = z^m\left[X(z) - \sum_{k=0}^{m-1} x(k)z^{-k}\right] \tag{3.20}$$

(This is the subject of Problem 3.9 at the end of the chapter.)

Specific examples, for $m = 1$ and $m = 2$, read, respectively,

$$\mathscr{Z}\{x(n+1)\} = z[X(z) - x(0)]$$

and

$$\mathscr{Z}\{x(n+2)\} = z^2[X(z) - x(0) - x(1)z^{-1}]$$

Damping, in the time domain, means we are considering sampled values of (say) $e^{-at}x(t)$. The Z-transform of the sampled sequence is, by definition,

$$\mathscr{Z}\{e^{-anT}x(nT)\} = x(0) + x(T)e^{-aT}z^{-1} + x(2T)e^{-2aT}z^{-2} + \cdots$$
$$= X(e^{+aT}z) \tag{3.21}$$

As usual, this last result can be invoked to establish the transform pairs for any function which can be expressed in terms of exponential functions.

The properties of the Z-transform are summarized very briefly in Table 3.1. The 'limit' theorems feature in Problem 3.11 at the end of the chapter.

3.12 Z-TRANSFORM PAIRS

In Section 3.9 we considered some elementary transforms, using the definition

$$X(z) = \sum_{0}^{\infty} x(nT)z^{-n}$$

as given by (3.13). The results obtained there are given as the first transform pairs of Table 3.2.

Worked Example 3.3

Obtain the Z-transform of
(i) $\cosh(anT)$ (see pair ZT5)
(ii) $\cos(anT)$ (see pair ZT6)

Table 3.2 Z-transform pairs

	Sampled time	Transform
ZT1	$x(nT) = \begin{cases} 1, & n=0 \\ 0, & n \neq 0 \end{cases}$	1
ZT2	$x(nT) = 1$	$1/(1 - z^{-1})$
ZT3	$e^{-anT}, a > 0$	$1/(1 - e^{-aT}z^{-1})$
ZT4	K^n	$1/(1 - Kz^{-1})$
ZT5	$\cosh(anT)$	$[(1 - z^{-1}\cosh(aT)]/[1 - 2z^{-1}\cosh(aT) + z^{-2}]$
ZT6	$\cos(anT)$	As ZT5, replacing 'cosh' by 'cos'
ZT7	$\sinh(anT)$	$z^{-1}\sinh(aT)/[1 - 2z^{-1}\cosh(aT) + z^{-2}]$
ZT8	$\sin(anT)$	As ZT7, replacing 'sinh' by 'sin' and 'cosh' by 'cos'
ZT9	nT	$Tz^{-1}/(1 - z^{-1})^2$
ZT10	$(nT)^2$	$T^2(z^{-1} + z^{-2})/(1 - z^{-1})^3$

Solution

In Worked Example 3.2(iii) we obtained the result

$$\mathscr{Z}\{e^{-anT}\} = (1 - e^{-aT}z^{-1})^{-1}$$

(i) Quoting that result, given the equation

$$\cosh(anT) = \tfrac{1}{2}(e^{anT} + e^{-anT})$$

we have

$$\mathscr{Z}\{\cosh(anT)\} = \tfrac{1}{2}[(1 - e^{aT}z^{-1})^{-1} + (1 - e^{-aT}z^{-1})^{-1}]$$

$$= \frac{1}{2}\left[\frac{2 - z^{-1}(e^{aT} + e^{-aT})}{1 - z^{-1}(e^{aT} + e^{-aT}) + z^{-2}}\right]$$

$$= \frac{[1 - z^{-1}\cosh(aT)]}{[1 - 2z^{-1}\cosh(aT) + z^{-2}]}$$

(ii) Similarly, we may write

$$\cos(anT) = \tfrac{1}{2}(e^{janT} + e^{-janT})$$

and come to the conclusion that

$$\mathscr{Z}\{\cos(anT)\} = \frac{[1 - z^{-1}\cos(aT)]}{[1 - 2z^{-1}\cos(aT) + z^{-2}]}$$

(We leave the above result for the reader to verify, using the same method.) ●

3.13 INVERSION

We have already noted that the formal inverse of the Z-transform is an integral in the complex plane of z. As it is *not* assumed that readers are familiar with methods of contour integration and the residue theorem, we shall look at other methods of inverting a known transform. Use of Table 3.2 is one possibility. This is only available, however, should $X(z)$ be the transform of a time-domain signal expressible in *closed form* (i.e. in terms of known elementary functions, $x(n)$).

In any application it is possible that a Z-transform is first presented in a form such as

$$X(z) = \frac{N(z^{-1})}{D(z^{-1})} \tag{3.22}$$

in which the numerator and denominator are polynomials in z^{-1}. (Here we are using the notation $X(z)$ in a quite general sense, not necessarily associated with the input to a system. In Section 3.10 we discussed how the structure shown in (3.22) might arise in the context of the transfer function $H(z)$, described by (3.16).) If (3.22) is not readily invertible from tables of Z-transform pairs and properties, we can consider as an alternative a *division* approach. Polynomials can be subjected to a long-division process as can ordinary numbers. By this method we can obtain a series for $X(z)$, term by term. The sampled value $x(nT)$ is, by definition, the coefficient of z^{-n} in the series, and so we may read off $x(0), x(T), x(2T), \ldots$, even though we probably *cannot* obtain $x(nT)$ in closed form (i.e. in terms of an arbitrary integer n).

Worked Example 3.4

By division, obtain the first three terms $x(n)$ of the sequence whose transform is

$$X(z) = \frac{1}{1 - 1.1z^{-1} + 0.1z^{-2}}$$

Solution

Following the usual procedure for long division, we obtain

$$
\begin{array}{r}
1 + 1.1z^{-1} + 1.11z^{-2} + \cdots \\
(1 - 1.1z^{-1} + 0.1z^{-2})\overline{\smash{\big)}\,1 } \\
\underline{1 - 1.1z^{-1} + 0.1z^{-2}} \\
1.1z^{-1} - 0.1z^{-2} \\
\underline{1.1z^{-1} - 1.21z^{-2} + 0.11z^{-3}} \\
1.11z^{-2} - 0.11z^{-3} \cdots
\end{array}
$$

So far we have $X(z)$ in series form, beginning

$$X(z) = 1 + 1.1z^{-1} + 1.11z^{-2} + \cdots$$

and the coefficient values give $x(0) = 1$, $x(T) = 1.1$ and $x(2T) = 1.11$, etc. ●

The preceding example is of particularly simple form, and can be handled differently. If the polynomial $D(z^{-1})$ has factors we can express $X(z)$ in terms of partial fractions and then refer to Tables 3.1

and 3.2 to identify the corresponding time-domain sequences. In the previous example, factorization gives

$$X(z) = \frac{1}{1 - 1.1z^{-1} + 0.1z^{-2}} = \frac{1}{(1 - z^{-1})(1 - 0.1z^{-1})}$$

from which

$$X(z) = \frac{1}{9}\left[\frac{10}{1 - z^{-1}} - \frac{1}{1 - 0.1z^{-1}} \right]$$

in partial fraction form. The constituent terms can be identified, by transform Z4, with the sequence

$$x(nT) = \tfrac{1}{9}[10(1)^n - (0.1)^n]$$

Putting $n = 0, 1, 2, 3, \ldots$, generates values $\{1, 1.1, 1.11, 1.111, \ldots\}$. The final limit is $\lim_{n \to \infty} x(nT) = \frac{10}{9}$, and we can confirm that this is $\lim_{z \to 1}(1 - z^{-1})X(z)$.

Similarly, if the denominator $D(z^{-1})$ is such that partial fractions of $X(z)$ include terms such as $(A + Bz^{-1})/(a + bz^{-1} + cz^{-2})$, we might be able to make use of the transform pairs for the trigonometric and hyperbolic sampled functions given by pairs ZT5–ZT8 of Table 3.2 (possibly using also the damping property Z2(ii) of Table 3.1).

However, if the quadratic has real coefficients but no real linear factors it will have roots which are complex conjugates and we still have the option of using *linear* partial fractions and again using the result $\mathscr{Z}\{(K)^n\} = (1 - Kz^{-1})^{-1}$. In this, K now would be a complex number, $re^{j\theta}$. After inversion, powers of $re^{j\theta}$ and its conjugate $re^{-j\theta}$ can then be combined and written as multiples of $\sin(n\theta)$ and $\cos(n\theta)$.

Such methods will be illustrated fully in Section 4.10, in the context of difference equations.

In conclusion, we observe that if the inverse of some transform $X(z)$ has been obtained, then we can invert expressions like $z^m X(z)$ and $z^{-m}X(z)$ using the delay properties Z1 of Table 3.1, which have the effect of shifting terms in the time domain through $\pm mT$. (The Laplace transform equivalent is the second shift theorem, property L2 of Table 1.1, in which inversion of a transform which has an exponential factor results in a shift in the time domain.) Use of the Z-transform delay properties also will be fully demonstrated in the next chapter, although these last points (about quadratics with complex roots, and delay properties) *do* appear in a simple example at the end of this chapter (Problem 3.10).

3.14 DISCRETE CONVOLUTION

To decide what is the time-domain interpretation of convolution, if this is to be identified with multiplication in the transform domain, we can begin by considering the product of two known Z-transforms, $X_1(z)$ and $X_2(z)$, and denote the transform of the convolution by $X(z)$. We need to find out what is the form of $X(z)$.

From the definition of the Z-transform as a series, (3.13), we have

$$
\begin{aligned}
X(z) &= x(0) + x(T)z^{-1} + x(2T)z^{-2} + x(3T)z^{-3} + \cdots \\
&= X_1(z)X_2(z) \\
&= \{x_1(0) + x_1(T)z^{-1} + x_1(2T)z^{-2} + \cdots\} \\
&\quad \times \{x_2(0) + x_2(T)z^{-1} + x_2(2T)z^{-2} + \cdots\}
\end{aligned} \tag{3.23}
$$

Term-by-term multiplication of the two series on the right-hand side of this equation allows comparison of the coefficients of successive powers of z^{-1}. Matching coefficients for the first few terms in the product reveals a pattern whereby the coefficient $x(nT)$ of the general power z^{-n} can be recognized.

We obtain, for $n = 0, 1, 2, 3$, the coefficients of z^0, z^{-1}, z^{-2} and z^{-3} as follows:

$$
\begin{aligned}
x(0) &= x_1(0)x_2(0) \\
x(T) &= x_1(0)x_2(T) + x_1(T)x_2(0) \\
x(2T) &= x_1(0)x_2(2T) + x_1(T)x_2(T) + x_1(2T)x_2(0) \\
x(3T) &= x_1(0)x_2(3T) + x_1(T)x_2(2T) + x_1(2T)x_2(T) + x_1(3T)x_2(0)
\end{aligned} \tag{3.24}
$$

We perceive that, for each value of n, the coefficient of z^{-n} in $X(z)$ can be written in the form

$$
x(nT) = \sum_{m=0}^{n} x_1(mT)x_2(nT - mT) \qquad n = 0, 1, 2, \ldots \tag{3.25}
$$

(The reader should verify that (3.25) *does* give the explicit results listed for the cases $n = 0, 1, 2, 3$ in (3.24).)

Equation (3.25) is the *convolution sum* of two sampled signals $x_{1s}(t)$ and $x_{2s}(t)$ in the time domain and appears as property Z4 in Table 3.1. It is the discrete equivalent of the analogue convolution integral

$$
x(t) = x_1 * x_2 = \int_{-\infty}^{\infty} x_1(t')x_2(t - t') \, dt'
$$

Integration has been replaced by summation; the role of the continuous variable t is assumed by the discrete integer variable n and the integration variable t' is replaced by the summation integer m. We also observe that the same concepts of folding and shifting as were used in Section 2.8, to assist in the graphical interpretation of analogue convolution, are available to assist in the formulation of point-by-point products required in the computation of $x(nT)$ as given by (3.25). For any particular integer n we can regard (3.25) as the scalar product of two vectors whose elements are sampled values of $x_1(t)$ and $x_2(t)$. We may sketch $x_1(mT)$ and $x_2(mT)$ as functions of the discrete integer variable m, and then fold $x_2(mT)$ to illustrate $x_2(-mT)$ and shift the latter to obtain $x_2(nT-mT)$. Examination of (3.24) shows clearly the effect of incrementing n by one unit: sampled values in the x_2-vector appear one term 'later' in the next convolution sum. This suggests that a matrix can be used to show, on successive rows, what factors $x_2(nT-mT)$ are to multiply which values of $x_1(mT)$ in forming the scalar products $x(0)$, $x(T)$, $x(2T)$,..., of (3.25).

Worked Example 3.5

Find the convolution of the discrete signals $x_1(nT)$ and $x_2(nT)$, where

$$x_1(nT) = 1, 2, 0, 2, 1 \quad \text{for} \quad n = 0, 1, 2, 3, 4$$

and

$$x_2(nT) = 2, 2, 2, 2, 1, 1 \quad \text{for} \quad n = 0, 1, \dots, 5$$

(Assume either is zero for other values of n.)

Solution

Without ambiguity we can set $T = 1$ and put

$$x(n) = x_1 * x_2 = \sum_{m=0}^{n} x_1(m) x_2(n-m) \quad n = 0, 1, 2, \dots$$

We begin by sketching $x_1(m)$, $x_2(m)$ and the folded and shifted vector $x_2(n-m)$ as functions of m, described by Fig. 3.25. In Fig. 3.25(d), $x_2(n-m)$ is shown as if $n = 2$.

From the diagrams Figs. 3.25(a) and 3.25(d) we see that if $n < 0$ and if $-5 + n > 4$ then $x(n) = 0$ because $x_2(n-m)$ has non-zero values only at points m where $x_1(m)$ is zero. We need consider only integers $n = 0$ to $n = 9$, for which one or more terms in the convolution sum are such that non-zero values of $x_1(m)$ and $x_2(n-m)$ coincide on the axis of m. In Fig. 3.26 we set up a matrix with ten rows corresponding to $n = 0, 1, \dots, 9$, and five columns (to reflect the fact that the 'fixed'

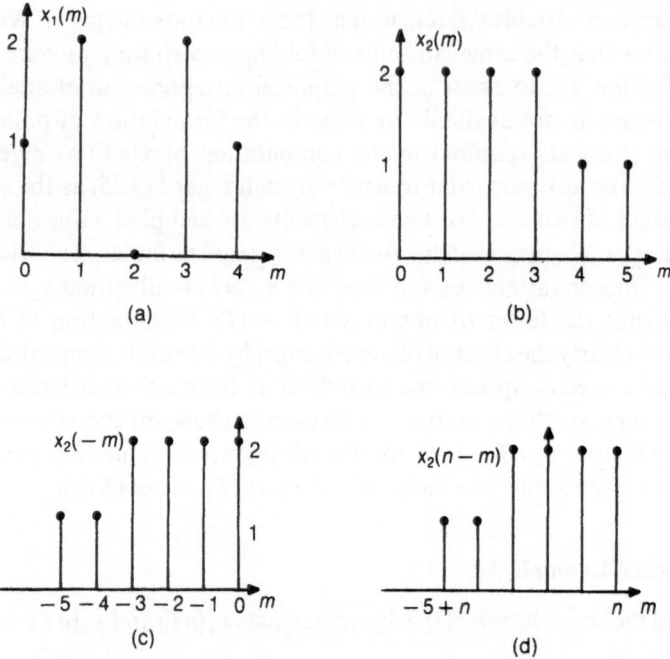

Fig. 3.25 Processes involved in the convolution $x_1 * x_2$.

m	0	1	2	3	4		$x(n)$
n	1	2	0	2	1	$\leftarrow x_1(m)$	
0	2					(1×2)	$= 2$
1	2	2				$(1 \times 2) + (2 \times 2)$	$= 6$
2	2	2	2			$(1 \times 2) + (2 \times 2) + (0)$	$= 6$
3	2	2	2	2		$(1 \times 2) + (2 \times 2) + (0) + (2 \times 2)$	$= 10$
4	1	2	2	2	2	$(1 \times 1) + (2 \times 2) + (0) + (2 \times 2) + (1 \times 2)$	$= 11$
5	1	1	2	2	2	$(1 \times 1) + (2 \times 1) + (0) + (2 \times 2) + (1 \times 2)$	$= 9$
6		1	1	2	2	$(2 \times 1) + (0) + (2 \times 2) + (1 \times 2)$	$= 8$
7			1	1	2	$(0) + (2 \times 2) + (1 \times 2)$	$= 4$
8				1	1	$(2 \times 2) + (1 \times 2)$	$= 3$
9					1	(1×1)	$= 1$

Fig. 3.26 Matrix showing the shifted positions of $x_2(n - m)$.

vector $x_1(m)$ has five elements. The values of $x_1(m)$ are shown immediately above the matrix. In relation to these, each row of the matrix on the left then shows where there are coinciding non-zero values of $x_2(n-m)$, for $n=0,1,\ldots,9$, with reference to Fig. 3.25(d). To the right of the matrix appears the corresponding scalar product, hence values of the convolution sum where $x(n)$ is non-zero. The convolution can be sketched as a function of n, if wished. We conclude from the last column of Fig. 3.26 that it has a Z-transform

$$X(z) = 2 + 6z^{-1} + 6z^{-2} + 10z^{-3} + 11z^{-4} + 9z^{-5} + 8z^{-6} + 4z^{-7} + 3z^{-8} + z^{-9} \qquad \bullet$$

Worked Example 3.6

Obtain $\mathcal{Z}\{x_1 * x_2\}$, where $x_1(n)$ and $x_2(n)$ are as defined in Worked Example 3.5, by direct evaluation of the product $X_1(z)X_2(z)$.

Solution

By definition, using the given data,

$$\begin{aligned} X_1(z)X_2(z) &= \{1 + 2z^{-1} + 2z^{-3} + z^{-4}\} \\ &\quad \times \{2 + 2z^{-1} + 2z^{-2} + 2z^{-3} + z^{-4} + z^{-5}\} \\ &= 2 + 6z^{-1} + 6z^{-2} + 10z^{-3} + 11z^{-4} + 9z^{-5} \\ &\quad + 8z^{-6} + 4z^{-7} + 3z^{-8} + z^{-9} \end{aligned}$$

which is the result obtained in Worked Example 3.5. $\qquad \bullet$

SUMMARY

In Chapter 3 we have considered the processes involved in obtaining a set of sampled values $x(nT)$ from an analogue signal $x(t)$, with particular reference to ideal sampling at regular intervals, and the consequential effects on the spectrum $X(f)$ – such as aliasing and induced periodicity. We have also considered the effects and implications of truncating summations by windowing sampled values in either the time or the transform domain.

Recognizing that discrete transform methods are required to deal with discrete data sets, we established the Z-transform as a discrete version of the Laplace transform, together with some of its most useful properties and a table of transform pairs. In conclusion, by considering the product of two Z-transforms, we obtained an expression, in the form of a sum, for the convolution of two time-domain sampled value sequences.

PROBLEMS

3.1 (i) Define $x_{nat} = (0, 1, 2, 3, 4, 5, 6, 7)^T$ to be the column vector which is a sequence of decimal numbers in natural order. Show that if each number is replaced by its binary equivalent, which is then bit-reversed, the resulting decimal vector is

$$x_{br} = (0, 4, 2, 6, 1, 5, 3, 7)^T$$

(ii) The symmetric permutation matrix E is defined to be

$$\begin{bmatrix} 1 & 0 & 0 & 0 & 0 & 0 & 0 & 0 \\ 0 & 0 & 0 & 0 & 1 & 0 & 0 & 0 \\ 0 & 0 & 1 & 0 & 0 & 0 & 0 & 0 \\ 0 & 0 & 0 & 0 & 0 & 0 & 1 & 0 \\ 0 & 1 & 0 & 0 & 0 & 0 & 0 & 0 \\ 0 & 0 & 0 & 0 & 0 & 1 & 0 & 0 \\ 0 & 0 & 0 & 1 & 0 & 0 & 0 & 0 \\ 0 & 0 & 0 & 0 & 0 & 0 & 0 & 1 \end{bmatrix}$$

Denoting $x(kT)$ by x_k, and defining

$$x_{nat} = (x_0, x_1, x_2, x_3, x_4, x_5, x_6, x_7)^T$$

show that

$$Ex_{nat} = (x_0, x_4, x_2, x_6, x_1, x_5, x_3, x_7)^T$$

(Note that the permutation matrix has the effect of reordering inputs so that they appear with the time suffix k bit-reversed. Such reordering is required in Chapter 7, in the context of developing fast methods of computing the discrete Fourier transform.)

3.2 If the sampling function $p(t)$ is the periodic pulse train

$$p(t) = \sum_{-\infty}^{\infty} \Pi\left(\frac{t - nT_s}{\tau}\right)$$

(as instanced in Section 3.3) show that, in (3.3), the weights C_n appearing in the transform $X_s(f)$ of the sampled function are given by

$$C_n = f_s \tau \, \text{sinc}(n f_s \tau)$$

3.3 (Parseval's theorem.) We may write (for an analogue signal),

$$\text{Energy} = \int_{-\infty}^{\infty} |x(t)|^2 \, dt = \int_{-\infty}^{\infty} x(t)\bar{x}(t) \, dt$$

where $\bar{x}(t)$ is the complex conjugate of $x(t)$.

From (1.13), substitute the inversion integral for $x(t)$ in this. By changing the order of integration in the resulting double integral, show that

$$\text{Energy} = \int_{-\infty}^{\infty} X(f)\bar{X}(f) \, df = \int_{-\infty}^{\infty} |X(f)|^2 \, df$$

(In terms of frequency $w = 2\pi f$ we would obtain

$$\text{Energy} = \frac{1}{2\pi} \int_{-\infty}^{\infty} |X(w)|^2 \, dw$$

3.4 In Worked Example 3.2(iv) we obtained the result

$$\mathscr{Z}\{nTe^{-anT}\} = \frac{Te^{-aT}z^{-1}}{(1 - e^{-aT}z^{-1})^2}$$

Using this result and similar methods obtain the transforms $\mathscr{Z}\{nT\}$ and $\mathscr{Z}\{(nT)^2\}$, given as pairs ZT9 and ZT10 of Table 3.2.

3.5 From the result

$$\mathscr{Z}\{e^{-anT}\} = 1/(1 - e^{-aT}z^{-1})$$

(pair ZT3), deduce the result $\mathscr{Z}\{K^n\}$ given in pair ZT4 of Table 3.2, wherein K is a constant.

3.6 Proceeding as in Worked Example 3.3, obtain the Z-transforms of

(i) $\sinh(anT)$
(ii) $\sin(anT)$

as given in pairs ZT7 and ZT8 of Table 3.2.

3.7 (i) By division, show that the first six terms of the sampled sequence whose transform is

$$X(z) = (1 - z^{-1})/(1 + z^{-3})$$

are as given in the following table (taking $T = 1$).

n	0	1	2	3	4	5	$\cdots$
$x(n)$	1	-1	0	-1	1	0	$\cdots$

(ii) Show that $X(z)$ as given in (i) above can be written in partial fraction form as

$$X(z) = \frac{1}{3}\left(\frac{2}{1 + z^{-1}} + \frac{1 - 2z^{-1}}{1 - z^{-1} + z^{-2}} \right)$$

On comparison of the denominators in this expression with those of pairs ZT4, ZT6 and ZT8 of Table 3.2, with what values of the constant K and the angle aT (with $T = 1$) can we associate the terms in $X(z)$? Write the second term in the form

$$\frac{1 - 2z^{-1}}{1 - z^{-1} + z^{-2}} = \lambda_1 \mathscr{Z}\{\cos(anT)\} + \lambda_2 \mathscr{Z}\{\sin(anT)\}$$

and by comparing coefficients show that $\lambda_1 = 1$ and $\lambda_2 = -\sqrt{3}$.

Hence show that the inverse of $X(z)$ can be expressed in closed form as the time-domain sequence

$$x(n) = \frac{2}{3}\left[(-1)^n + \cos\left[(n + 1)\frac{\pi}{3} \right] \right]$$

and verify that this generates the values obtained in part (i) of this question.

3.8 (i) From the result

$$\mathscr{Z}\{x(nT)e^{-anT}\} = X(e^{aT}z)$$

show that

$$\mathscr{Z}\{x(nT)\alpha^{-nT}\} = X(\alpha^T z)$$

(property Z2(ii) of Table 3.1).

(ii) Show that

$$\mathscr{Z}\left\{2^{-n}\cos\left(\frac{n\pi}{4}\right)\right\} = \left(1 - \frac{1}{2\sqrt{2}}z^{-1}\right)\Big/\left(1 - \frac{1}{\sqrt{2}}z^{-1} + \frac{1}{4}z^{-2}\right)$$

3.9 Premultiplication of the defining sum of the Z-transform by z^m gives the equation

$$z^m X(z) = \sum_{0}^{\infty} x(n)z^{-n+m}$$

Proceeding in a manner similarly to that used in Section 3.11 (including appropriate changes of summation integers) show that

$$z^m X(z) = \mathscr{L}\{x(p+m)\} + \sum_{p=-m}^{-1} x(p+m)z^{-p}$$

and that this can be written

$$z^m X(z) = \mathscr{L}\{x(p+m)\} + \sum_{k=0}^{m-1} x(k)z^{-(k-m)}$$

(This proves the delay property Z1(ii) of Table 3.1 if p is now replaced by n.)

3.10 (i) From Table 3.2 show that

$$\mathscr{L}^{-1}\left\{\frac{z^{-1}}{1+z^{-2}}\right\} = \sin\left(n\frac{\pi}{2}\right)$$

(ii) Show that, in partial fraction form

$$\frac{z^{-1}}{1+z^{-2}} = \frac{1}{2j}\left[\frac{1}{1-jz^{-1}} - \frac{1}{1+jz^{-1}}\right]$$

Express the numbers $\pm j$ in the form $re^{j\theta}$. Hence show that use of an alternative transform pair leads to the same result as in (i) above.

(iii) Use an appropriate delay property to show that

$$\mathscr{L}^{-1}\left\{\frac{1}{1+z^{-2}}\right\} = \sin\left[(n+1)\frac{\pi}{2}\right]$$

3.11 (i) Show that the initial value theorem

$$\mathscr{L}\{x(0)\} = \lim_{z\to\infty} X(z)$$

is a direct consequence of the defining sum for $\mathscr{L}\{x(nT)\}$.

(ii) Use a delay property to show that

$$(1-z^{-1})\sum_{0}^{N} x(nT)z^{-n} = \sum_{0}^{N}\{x(nT) - x(nT-T)\}z^{-n}$$

(Assume $x(-T) = 0$.) By setting $z = 1$ on the right-hand side, obtain a rough justification of the final value theorem. (See property Z3 of Table 3.1.)

3.12 Define

$$x_1(nT) = \cos\left(\frac{n\pi}{3}\right) \quad \text{for} \quad n = 0, 1, 2, 3, 4$$

$$(x_1(nT) = 0 \text{ for } n < 0, n > 4)$$

and

$$x_2(nT) = \sin\left(\frac{n\pi}{3}\right) \quad \text{for} \quad n = 0, 1, 2, 3$$

$$(x_2(nT) = 0 \text{ for } n < 0, n > 3)$$

Use the method described in Section 3.14 to find the convolution $x_1 * x_2$, and deduce that

$$Z\{x_1 * x_2\} = \tfrac{\sqrt{3}}{2}[z^{-1} + \tfrac{3}{2}z^{-2} - \tfrac{3}{2}z^{-4} - \tfrac{3}{2}z^{-5} - \tfrac{1}{2}z^{-6}]$$

Confirm this result by direct evaluation of the product $X_1(z)X_2(z)$.

4

Difference equations and the Z-transform

The context in which difference equations might appear as discrete versions of differential equations has already been instanced in Section 3.10, where we considered the digital description of the transfer function of a linear input–output system. Difference equations, however, might arise directly – for example, in the description of ladder networks or the distribution of bending moment along a load-bearing beam supported at a number of separated points.

In this chapter, we shall first consider difference equations in the context of approximations to differential equations, and then refer briefly to systems modelled directly by difference equations such as those to which we have just referred. Most of the chapter is then concerned with the application of the Z-transform to the solution of difference equations, using the properties and pairs established in Chapter 3. When considering particular examples, we shall illustrate various methods of approach: the *ad hoc* (or trial) method of solution; an approach to inversion which involves only partial fractions and the subsequent use of tables; and, finally, methods of inversion which invoke the delay properties of the Z-transform and (possibly) the convolution theorem.

The comparison between methods used in the solution of differential and difference equations can again be emphasized. Trial methods used in the solution of linear differential equations with constant coefficients involve postulating an exponential form of solution of the homogeneous equation (i.e. of the complementary function) and also a form of particular integral related to the input (or 'forcing') terms seen in the inhomogeneous equation. Similarly, it might be possible to anticipate the form of solution of a difference equation (this will be illustrated in Worked Example 4.1). We note that trial methods of solution of problems (whether continuous or discrete) do not involve transformations. The analogy continues when the describing equation *is* first transformed. In the continuous-time problem, solution requires the inversion of a Laplace transform, which might be possible using only partial fractions and tables, or might involve also the Laplace transform shift (and other) properties. It has been indicated already that equivalent methods are available when we need to invert a Z-transform.

4.1 FORWARD AND BACKWARD DIFFERENCE OPERATORS

The *forward* difference operator Δ is defined by the equation

$$\Delta y_n = y_{n+1} - y_n \tag{4.1}$$

Higher-order operators are then defined inductively. For example,

$$\Delta^2 y_n = \Delta(\Delta y_n) = \Delta(y_{n+1} - y_n) = \Delta y_{n+1} - \Delta y_n$$
$$= (y_{n+2} - y_{n+1}) - (y_{n+1} - y_n)$$
$$= y_{n+2} - 2y_{n+1} + y_n$$

Similarly,

$$\Delta^3 y_n = y_{n+3} - 3y_{n+2} + 3y_{n+1} - y_n$$

Note that the coefficients of the terms arising in $\Delta^k y_n$ are those which appear in the binomial expansion of $(1 - x)^k$.

The *backward* difference operator is defined in like manner by the equation

$$\Delta y_n = y_n - y_{n-1} \tag{4.2}$$

from which we can obtain the results

$$\Delta^2 y_n = y_n - 2y_{n-1} + y_{n-2}$$
$$\Delta^3 y_n = y_n - 3y_{n-1} + 3y_{n-2} - y_{n-3}$$

which have the same structure as those for forward differences, but now $\Delta^k y_n$ is expressed in terms of y_n and *previous* values in the sequence.

4.2 THE APPROXIMATION OF A DIFFERENTIAL EQUATION

There are various ways of approximating the derivatives of a continuous variable $y(t)$. At the simplest level one could put either

$$\frac{dy}{dt} = \frac{y_{n+1} - y_n}{T}$$

or

$$\frac{dy}{dt} = \frac{y_n - y_{n-1}}{T}$$

at $t = nT$, where T is the interval between successive sequence values. More generally, an expression approximating $d^k y/dt^k$ is some weighted combination of sampled values of y at and in the vicinity of the 'point' $t = nT$. Whatever approximation methods are selected, a linear differential equation of degree N will be replaced by an expression which can be written

$$a_0 \Delta^N y_n + a_1 \Delta^{N-1} y_n + \cdots + a_n y_n = r_n \qquad a_0 \neq 0$$

in which the coefficients $\{a_k\}$ are known, and r_n is a sequence of sampled values of the input (or forcing) term of the differential equation. If $r_n \not\equiv 0$, this is an inhomogeneous difference equation, otherwise the equation is homogeneous in y_n.

For example, the equation describing an input–output system, (3.15), can be rewritten in simplified notation in the following form:

$$y_n + a_1 y_{n-1} + a_2 y_{n-2} + \cdots + a_q y_{n-q} = b_0 x_n + b_1 x_{n-1} + \cdots + b_p x_{n-p}$$

This is a backward difference equation of order $N = q$. The order of a difference equation, in other words, is the same as the number of previous outputs appearing. (Similarly, a difference equation containing terms $y_n, y_{n+1}, \ldots, y_{n+m}$ is of order $N = m$.)

A difference equation, we see, can be regarded as a recurrence equation, and if it is of order N then, in principle, we need the values of N terms in order to calculate the next term in the sequence. In practice, information about the solution might include a requirement that y_n be finite as $n \to \infty$. The situation is analogous to the fact that we need N initial and/or boundary conditions to obtain a particular solution of an Nth-order differential equation.

4.3 LADDER NETWORKS

Applying Kirchhoff's laws to model the behaviour of the nth mesh of a ladder network, depending on the components in the mesh, might result in a description which involves *simultaneous* difference

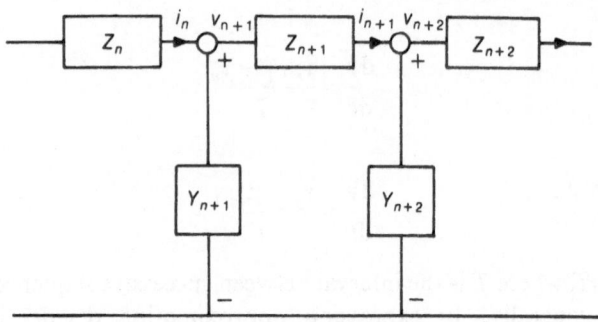

Fig. 4.1 A mesh in a ladder network.

equations. For example, the difference equations describing the network illustrated in Fig. 4.1 take the following form:

$$v_{n+1} = i_{n+1}Z_{n+1} + v_{n+2}$$
$$i_n = v_{n+1}Y_{n+1} + i_{n+1}$$

in which Z and Y are known impedances and admittances. To find the mesh voltages and currents v and i would in this case necessitate solution of simultaneous difference equations subject to given end conditions.

As is the case when applying the Laplace transform, to solve simultaneous differential equations, after transforming using Z-transform properties we would obtain (in this example) simultaneous equations in, say, $V(z)$ and $I(z)$ between which we may eliminate one transform and invert the resulting expression of the other. The analogy is straightforward and we shall concentrate in this chapter on methods used for just *one* difference equation.

4.4 BENDING IN BEAMS: TRIAL METHODS OF SOLUTION

In Problem 1.12 was given a fourth-order equation describing the deflection $y(x)$ of a horizontal beam, subject to a point load. (The boundary conditions given there, namely $y = dy/dx = 0$, were those appropriate when ends are clamped horizontally.) That equation is a particular example of the more general equation

$$\frac{d^4y}{dx^4} = \frac{w(x)}{EI}$$

in which $w(x)$ is a function describing a prescribed distributed load, E is Young's modulus of elasticity, and I is the moment of inertia

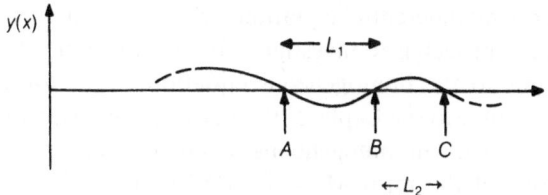

Fig. 4.2 A uniformly loaded, simply supported, beam.

about the horizontal axis $0x$. If the beam is uniform, in terms of bending moment M we can write

$$\frac{d^2M}{dx^2} = w(x), \quad \frac{d^2y}{dx^2} = \frac{M}{EI}$$

(At any point, bending moment $M = EI/R$, where R is the radius of curvature of the centre-line of.the beam.) Derivation of these results can be found in any basic text on statics, as can the following theorem of three moments, which is applicable when the beam is uniformly loaded ($w(x) = $ constant), and is simply supported at a number of points at the same level, as shown in Fig. 4.2.

In Fig. 4.2 are shown just three supports, at A, B and C, but these might be any three of several neighbouring supports. By 'simply supported' is meant 'pin-jointed', and there is no prescribed gradient dy/dx at the point of support. In this case, bending moment values at A, B and C are related by Clapeyron's theorem of three moments, which is stated by the equation

$$L_2 M_C + 2(L_1 + L_2)M_B + L_1 M_A = w(L_1^3 + L_2^3)/4 \qquad (4.3)$$

in which L_1 and L_2 are the distances between the successive supports.

Worked Example 4.1

Consider a uniformly loaded long beam, $x \geqslant 0$, simply supported at equidistant points $x = nL$. Find the bending moments at these points, given that $M_0 = 0$ and $\lim_{n \to \infty} M_n$ is finite.

Solution

Equation (4.3) can be replaced by a difference equation

$$M_{n+2} + 4M_{n+1} + M_n = wL^2/2 \qquad (4.4)$$

(The right-hand-side sequence, $\{r_n\}$, is now constant.)

When solving differential equations with constant coefficients, an elementary approach is to investigate, by substitution, whether there are solutions to the homogeneous equation of the form $y = Be^{\lambda t}$, (without recourse to the Laplace transform). An equivalent approach when dealing with inhomogeneous difference equations is to look for solutions of the form $M_n = A + BK^n$ (without recourse to the Z-transform). In this trial solution, we can regard BK^n as the equivalent of a complementary function, and constant A as the equivalent of a particular integral, having noted that $wL^2/2$ is constant.

If (4.4) has a solution with this structure, then

$$(A + BK^{n+2}) + 4(A + BK^{n+1}) + (A + BK^n) = wL^2/2$$

By comparison, we obtain the equations $6A = wL^2/2$ and $BK^n(K^2 + 4K + 1) = 0$. We reject the possibility that $B = 0$ or $K = 0$ (a trivial solution which would not allow us to satisfy the condition $M_0 = 0$). Solving the quadratic gives $K = -2 \pm \sqrt{3}$. Powers of $K = -2 - \sqrt{3}$ oscillate between increasingly large positive and negative values and so, given that M_n is to be finite as $n \to \infty$, we accept only $K = -2 + \sqrt{3}$, for which $|K| < 1$. The trial solution reduces to

$$M_n = \frac{wL^2}{12} + B(-2 + \sqrt{3})^n$$

Since $M_0 = 0$ we obtain $B = -wL^2/12$ to complete the solution, i.e. bending moments at points $x = nL$ are given by

$$M_n = wL^2\{1 - (-2 + \sqrt{3})^n\}/12 \qquad \bullet$$

Trial methods can be modified to deal with difference equations which are not so straightforward (as can trial methods for finding the complementary function and a particular integral of a differential equation), but we shall develop more systematic methods, based on the Z-transform.

In concluding this section, we notice that the nearest electrical equivalent of the problem of Worked Example 4.1 is a ladder network in which meshes have equal impedances.

4.5 TRANSFORMING A SECOND-ORDER FORWARD DIFFERENCE EQUATION

In general terms we might write the second-order forward difference equation as

$$y_{n+2} + c_1 y_{n+1} + c_0 y_n = r_n \qquad (4.5)$$

(of which (4.4) was an example). It is assumed that r_n is a known sequence (finite or infinite) and that values y_0 and y_1 are known.

Using the delay property Z1(ii) of Table 3.1, with $m = 2$ and $m = 1$, (4.5) can be rewritten in terms of the Z-transform $Y(z) = \mathscr{Z}\{y_n\}$ as

$$z^2[Y(z) - y_0 - y_1 z^{-1}] + c_1 z[Y(z) - y_0] + c_0 Y(z) = R(z)$$

from which

$$Y(z) = \frac{1}{z^2 + c_1 z + c_0} [z^2 y_0 + z(y_1 + c_1 y_0) + R(z)] \qquad (4.6)$$

The solution of (4.5) is the inverse of this expression. Looking at the first and third terms on the right-hand side of (4.6), we note that

$$\frac{z^2}{z^2 + c_1 z + c_0} = \frac{z(z)}{z^2 + c_1 z + c_0} \qquad (4.7)$$

and

$$\frac{R(z)}{z^2 + c_1 z + c_0} = \frac{z^{-1}(z)}{z^2 + c_1 z + c_0} R(z) \qquad (4.8)$$

This suggests that if we invert the second term in (4.6), which is a constant multiple of

$$\frac{z}{z^2 + c_1 z + c_0} = \frac{z^{-1}}{1 + c_1 z^{-1} + c_0 z^{-2}}$$

we can invoke *both* delay properties and (in the case of the term described in (4.8)) the convolution property of the Z-transform to complete the solution. (Depending on the form of the sequence r_n, however, it might be more straightforward to use methods other than convolution.)

4.6 THE CHARACTERISTIC POLYNOMIAL AND THE TERMS TO BE INVERTED

The divisor $(z^2 + c_1 z + c_0)$ appearing in (4.6) is usually referred to as the *characteristic polynomial* of the transformed equation, written

$$p(z) = (z^2 + c_1 z + c_0) \qquad (4.9)$$

and this, obviously, is the denominator of $Y(z)$ in the transformed problem. The roots of the equation $p(z) = 0$ might be real and distinct, or complex, or we might find the roots are equal. (Subsequent remarks can be generalized when the order of the equation exceeds $N = 2$.)

On division of both numerator and denominator by z^{-2}, the second term of (4.6) is a multiple of

$$\frac{z}{p(z)} = \frac{z^{-1}}{1 + c_1 z^{-1} + c_0 z^{-2}}$$

Suppose that this can be identified as the transform of a time-domain sequence $\tilde{y}_n$, so that

$$\mathscr{Z}^{-1}\left\{\frac{z}{p(z)}\right\} = \tilde{y}_n \tag{4.10}$$

Sections 4.7, 4.8 and 4.9 deal with examples in which the roots of $p(z) = 0$ are respectively real, complex or repeated, and it will be seen in each case that, when the sequence $\tilde{y}_n$ of (4.10) has been identified, we find $\tilde{y}_0 = 0$. That result will be anticipated as we complete this general discussion of the solution process.

As suggested in Section 4.5, we can next invert the term $z^2/p(z)$ using delay property Z1(ii) of Table 3.1 with $m = 1$. That can be written in the form

$$\mathscr{Z}\{\tilde{y}_{n+1}\} = z[\mathscr{Z}\{\tilde{y}_n\} - \tilde{y}_0]$$

and putting $\tilde{y}_0 = 0$ we have

$$\mathscr{Z}\{\tilde{y}_{n+1}\} = z\frac{z}{p(z)}$$

so that

$$\mathscr{Z}^{-1}\left\{\frac{z^2}{p(z)}\right\} = \tilde{y}_{n+1} \tag{4.11}$$

Similarly, we may use delay property Z1(i) with $m = 1$ in the form

$$\mathscr{Z}\{\tilde{y}_{n-1}\} = z^{-1}[\mathscr{Z}\{\tilde{y}_n\} + z\tilde{y}_{-1}]$$

As we are using forward differences and assuming that $y_n = 0$ for $n < 0$, it is assumed that any constituent term $\tilde{y}_n = 0$ for $n < 0$. The previous equation then reduces to

$$\mathscr{Z}\{\tilde{y}_{n-1}\} = z^{-1}\frac{z}{p(z)}$$

giving the result

$$\mathscr{L}^{-1}\left\{\frac{1}{p(z)}\right\} = \tilde{y}_{n-1} \tag{4.12}$$

We would use (4.12) if we wished to invert the term $R(z)/p(z)$ using the convolution property Z4 of Table 3.1, in which case the inverse of $1/p(z)$ has to be identified.

4.7 THE CASE WHEN THE CHARACTERISTIC POLYNOMIAL HAS REAL ROOTS

Suppose that $p(z)$, as defined in (4.9), has real roots. Then there are linear factors $p(z) = (z - K_1)(z - K_2)$ in which K_1 and K_2 are real and, for present purposes, assumed to be distinct, $K_1 \neq K_2$.

Consider first the term $z/p(z)$. We can obtain partial fractions of this in the form

$$\frac{z}{p(z)} = \frac{z^{-1}}{(1 - K_1 z^{-1})(1 - K_2 z^{-1})}$$

$$= \frac{1}{K_1 - K_2}\left(\frac{1}{1 - K_1 z^{-1}} - \frac{1}{1 - K_2 z^{-1}}\right) \tag{4.13}$$

With reference to (4.10) and pair ZT4 of Table 3.2, we identify this with the time-domain sequence

$$\tilde{y}_n = \frac{1}{K_1 - K_2}[(K_1)^n - (K_2)^n] \tag{4.14}$$

(In most texts $\tilde{y}_n$ is referred to as the *original sequence*, from which – as described in the previous section – other terms in the transfer $Y(z)$ can be inverted, using delay and convolution properties.)

Note that on substituting $n = 0$ in (4.14) we obtain $\tilde{y}_0 = 0$.

Multiplied by the known constant $(y_1 + c_1 y_0)$, (4.14) provides the inverse of the second term on the right-hand side of (4.6).

It follows from (4.11) that

$$\mathscr{L}^{-1}\left\{\frac{z^2}{p(z)}\right\} = \tilde{y}_{n+1}$$

and, from (4.12), we have

$$\mathscr{L}^{-1}\left\{\frac{1}{p(z)}\right\} = \tilde{y}_{n-1} \qquad n \geqslant 2$$

(In putting $n \geqslant 2$, we recall that $\tilde{y}_{-1} = \tilde{y}_0 = 0$.) With the appropriate change of suffix, the convolution property gives the result

$$\mathscr{Z}^{-1}\left\{\frac{1}{p(z)}R(z)\right\} = \sum_{m=2}^{n} \tilde{y}_{m-1}r_{n-m} \tag{4.15}$$

(If explanation is required, $\tilde{y}_{m-1}$ replaces $\tilde{y}_m$, because we have just used a delay property. Also, since $\tilde{y}_{-1} = 0$ and $\tilde{y}_0 = 0$, the lower summation limit becomes $m = 2$.)

If these results are combined, then the sequence $\{y_n\}$ as described by the difference equation (4.5) is given by the equation

$$y_n = y_0\tilde{y}_{n+1} + (y_1 + c_1 y_0)\tilde{y}_n + \sum_{2}^{n} \tilde{y}_{m-1}r_{n-m}$$

in which $\tilde{y}_n$ is given by (4.14).

Worked Example 4.2

Given the difference equation

$$y_{n+2} - 4y_{n+1} + 3y_n = 0$$

and arbitrary values y_0 and y_1, generate the sequence $\{y_n\}$ in terms of y_0 and y_1.

Solution

Transforming this equation as in Section 4.5, we obtain

$$z^2[Y(z) - y_0 - y_1 z^{-1}] - 4z[Y(z) - y_0] + 3Y(z) = 0$$

from which the characteristic polynomial is identified as

$$p(z) = z^2 - 4z + 3 = (z - 1)(z - 3)$$

whose roots are distinct and real; $K_1 = 1$ and $K_2 = 3$. We need to invert the transform

$$Y(z) = \frac{1}{p(z)}[z^2 y_0 + z(-4y_0 + y_1)]$$

which should be compared with (4.6).

Following the procedure described, we first invert the term

$$\frac{z}{(z-1)(z-3)} = -\frac{1}{2}\left\{\frac{1}{1-z^{-1}} - \frac{1}{1-3z^{-1}}\right\}$$

This is the equivalent of (4.13), and hence the original sequence is, from (4.14),

$$\tilde{y}_n = -\tfrac{1}{2}[1^n - 3^n] \qquad (\tilde{y}_0 = 0)$$

The term $z^2/(z^2 - 4z + 3)$ then inverts to give

$$\tilde{y}_{n+1} = -\tfrac{1}{2}(1^{n+1} - 3^{n+1})$$

and, as the equation presented is homogeneous, we have no further terms to consider. Therefore, inversion of $Y(z)$ gives the solution

$$y_n = -\tfrac{1}{2}y_0[1^{n+1} - 3^{n+1}] - \tfrac{1}{2}[-4y_0 + y_1][1^n - 3^n]$$

or

$$y_n = \tfrac{1}{2}[3y_0 - y_1] + \frac{3^n}{2}[-y_0 + y_1]$$

(The reader should confirm that this correctly reproduces initial values y_0 and y_1.)   ●

4.8 THE CASE WHEN THE CHARACTERISTIC POLYNOMIAL HAS COMPLEX ROOTS

Assume that the coefficients c_1 and c_0 in the characteristic polynomial $p(z) = z^2 + c_1 z + c_0$ are real. In that case, if the roots of $p(z) = 0$ are not real (whether distinct or not) they will be complex conjugate numbers. In other words, $K_1 = \alpha + j\beta$ and $K_2 = \alpha - j\beta$.

In Section 3.13 (and Problem 3.10(ii)) it was indicated that methods applicable to real linear factors can be extended to the present situation, as an alternative to making use of such transform pairs in Table 3.2 as have quadratic denominators.

The partial fractions of $z/p(z)$, as given in general form by (4.13), are now

$$\frac{z}{p(z)} = \frac{1}{2j\beta}\left[\frac{1}{1 - (\alpha + j\beta)z^{-1}} - \frac{1}{1 - (\alpha - j\beta)z^{-1}}\right]$$

Converting $K_1 = \alpha + j\beta$ to exponential form $re^{j\theta}$, (so that $K_1 = \alpha - j\beta = re^{-j\theta}$), the inverse of this is then

$$\tilde{y}_n = \frac{1}{2j\beta}[(re^{j\theta})^n - (re^{-j\theta})^n] = \frac{r^n}{\beta}\left[\frac{e^{jn\theta} - e^{-jn\theta}}{2j}\right] = \frac{r^n}{\beta}\sin(n\theta)$$

Delay properties can then be applied to complete the solution of a difference equation as described in Section 4.7, noting that $\tilde{y}_0 = 0$.

Worked Example 4.3

Given initial values $y_0 = \frac{1}{2}$ and $y_1 = 3$, find the sequence $\{y_n\}$ which satisfies the forward difference equation

$$y_{n+2} - 4y_{n+1} + 5y_n = 0$$

Solution

The transformed equation is

$$z^2[Y(z) - y(0) - y(1)z^{-1}] - 4z[Y(z) - y(0)] + 5Y(z) = 0$$

giving, after substituting the known values of y_0 and y_1,

$$Y(z) = \frac{1}{z^2 - 4z + 5}[\tfrac{1}{2}z^2 + (3 - 2)z]$$

To invert $z/p(z) = z^{-1}/(1 - 4z^{-1} + 5z^{-2})$ we can either use the damping property Z2(ii) of Table 3.1 in conjunction with the sine transform of Table 3.2, or note that $p(z) = 0$ has roots $2 \pm j$ and proceed as described above.

Using the first method, we begin by noting that

$$\mathscr{Z}\{\sin(an)\} = \frac{z^{-1}\sin a}{(1 - 2z^{-1}\cos a + z^{-2})}$$

and so, using the damping property,

$$\mathscr{Z}\{\alpha^{-n}\sin(an)\} = \frac{\alpha^{-1}z^{-1}\sin a}{1 - 2\alpha^{-1}z^{-1}\cos a + \alpha^{-2}z^{-2}}$$

Comparing the denominator with $(1 - 4z^{-1} + 5z^{-2})$, we choose $\alpha^{-2} = 5$, and $\alpha^{-1}\cos a = 2$, so that $\cos a = 2/\sqrt{5}$, whence $\sin a = 1/\sqrt{5}$. It follows that

$$\mathscr{Z}^{-1}\left\{\frac{z}{p(z)}\right\} = \tilde{y}_n = 5^{n/2}\sin(na)$$

in which a is a known acute angle. From (4.11), noting that $\tilde{y}_0 = 0$, it follows that

$$\mathscr{Z}^{-1}\left\{\frac{z^2}{p(z)}\right\} = \tilde{y}_{n+1} = 5^{(n+1)/2}\sin[(n+1)a]$$

The solution of the difference equation is then

$$y_n = \tfrac{1}{2}5^{(n+1)/2}\sin[(n+1)a] + 5^{n/2}\sin(na)$$

which can be re-expressed as

$$y_n = 5^{n/2}\{2\sin(na) + \tfrac{1}{2}\cos(na)\}$$

Using the second method, we begin by noting that the roots of $p(z) = 0$ can be written in exponential form as $K_1 = \sqrt{5}e^{ja}$ and $K_2 = \sqrt{5}e^{-ja}$ in which a is the acute angle $\tan^{-1}\tfrac{1}{2}$. The partial fraction form of $z/p(z)$ is then

$$\frac{z^{-1}}{1 - 4z^{-1} + 5z^{-2}} = \frac{1}{2j}\left(\frac{1}{1 - \sqrt{5}\,e^{ja}z^{-1}} - \frac{1}{1 - \sqrt{5}\,e^{-ja}z^{-1}}\right)$$

The inverse of this gives the original sequence in the form

$$\tilde{y}_n = \frac{1}{2j}[(\sqrt{5}\,e^{ja})^n - (\sqrt{5}\,e^{-ja})^n] = 5^{n/2}\sin(na)$$

The solution is completed as before.   ●

Worked Example 4.4

Find the sequence satisfying the forward difference equation

$$y_{n+2} - \sqrt{2}y_{n+1} + y_n = r_n$$

given that $y_0 = 1$, $y_1 = 0$ and that $r_0 = 2, r_1 = 1$ and $r_n = 0$ otherwise.

Solution

Proceeding as usual we obtain the transform of the sequence y_n, as in (4.6), in the form

$$Y(z) = \frac{1}{z^2 - \sqrt{2}z + 1}[z^2 y_0 + z(y_1 - \sqrt{2}y_0) + R(z)]$$

and insertion of initial values gives

$$\cdot \quad Y(z) = \frac{1}{z^2 - \sqrt{2}z + 1}[z^2 - \sqrt{2}z + R(z)]$$

(Note that we have *not* put $R(z) = 2 + z^{-1}$. Although the sequence $\{r_n\}$ and its transform in this example are very simple, we intend to take the opportunity to illustrate the use of convolution, as an alternative to inversion using partial fractions, tables and delay properties only.)

First consider the term in $z/p(z)$. The roots of $z^2 - \sqrt{2}z + 1 = 0$ are $K = (1 \pm j)/\sqrt{2}$ – complex conjugates. As explained at the beginning of this section, we can proceed as in the case of real roots, noting that $K_1 = e^{j\pi/4}$ and $K_2 = e^{-j\pi/4}$. Partial fractions, from (4.13), are given by

$$\frac{z}{p(z)} = \frac{z^{-1}}{1 - \sqrt{2}z^{-1} + z^{-2}} = \frac{1}{\sqrt{2}j} \left(\frac{1}{1 - e^{j\pi/4}z^{-1}} - \frac{1}{1 - e^{-j\pi/4}z^{-1}} \right)$$

whence

$$\tilde{y}_n = \frac{1}{\sqrt{2}j} (e^{j\pi n/4} - e^{-j\pi n/4}) = \sqrt{2} \sin\left(\frac{n\pi}{4}\right), \qquad \tilde{y}_0 = 0$$

(The same result is obtained by putting $a = \pi/4$ in the transform ZT8 of Table 3.2. No scaling is necessary in this case, as in the quadratic $(1 - \sqrt{2}z^{-1} + z^{-2})$ the coefficient of z^{-2} is already equal to 1.)

We know that

$$\mathscr{Z}^{-1}\left\{\frac{z^2}{p(z)}\right\} = \tilde{y}_{n+1}$$

from (4.11), and from (4.12) it follows that

$$\mathscr{Z}^{-1}\left\{\frac{1}{p(z)}\right\} = \tilde{y}_{n-1} \qquad \text{for } n \geqslant 2$$

If we choose to invert $R(z)/p(z)$ by convolution, (4.15) gives

$$\mathscr{Z}^{-1}\left\{\frac{R(z)}{p(z)}\right\} = \sum_{m=2}^{n} \tilde{y}_{m-1} r_{n-m} \qquad n \geqslant 2$$

At this point we pause to consider what terms are present in this convolution sum. There are only two non-zero terms, and these correspond to factors

$$r_{n-m} = r_0 = 2 \qquad \text{(signifying } m = n)$$

and

$$r_{n-m} = r_1 = 1 \qquad \text{(signifying } m = n - 1)$$

Substituting the appropriate values of m in the factor $\tilde{y}_{m-1}$, it follows that

$$\mathscr{Z}^{-1}\left\{\frac{R(z)}{p(z)}\right\} = 2\tilde{y}_{n-1} + \tilde{y}_{n-2} \qquad n \geqslant 2$$

Putting the three results together, with appropriate constant multipliers, the solution can be written

$$y_n = \tilde{y}_{n+1} - \sqrt{2}y_n + H(n-2)(2\tilde{y}_{n-1} + \tilde{y}_{n-2})$$

The inclusion of a step function indicates that the input $\{r_n\}$ does not affect values of the sequence before the term y_2 (as the difference equation itself indicated).

The reader is invited to verify that the equation

$$y_n = \sqrt{2}\left\{ \sin\left[\frac{(n+1)\pi}{4}\right] - \sqrt{2}\sin\left(\frac{n\pi}{4}\right) \right.$$
$$\left. + H(n-2)\left(2\sin\left[\frac{(n-1)\pi}{4}\right] + \sin\left[\frac{(n-2)\pi}{4}\right] \right) \right\}$$

correctly reproduces the values of y_0 and y_1, and gives values of y_2 and y_3 as could be obtained directly from the difference equation. ●

4.9 THE CASE WHEN THE CHARACTERISTIC EQUATION HAS REPEATED ROOTS

Clearly, if $p(z)$ is a quadratic which is a perfect square then there is no question of applying the partial fraction techniques discussed previously; and equally, we can expect that inversion in terms of hyperbolic or trigonometric sequences is not available in this special case.

In Worked Example 3.2(iv), and transforms ZT9 and ZT10 of Table 3.2 (the subject of Problem 3.4), we saw that the appearance of a repeated linear factor in the denominator of a transform can be associated with factors n in a time-domain signal. Differentiation of both signal and transform with respect to an appropriate quantity results in a transform denominator of higher power (and that is equally the case when there is a quadratic factor in $p(z)$). Those results are used in any case where we find repeated roots, and an example is sufficient to explain how to proceed.

Worked Example 4.5

Solve the second-order forward difference equation

$$4y_{n+2} - 4y_{n+1} + y_n = r_n$$

given that $y_0 = 1$, $y_1 = 3$ and that $r_0 = r_1 = 4$ and $r_n = 0$ otherwise.

Solution

Transforming the equation, we have

$$4z^2[Y(z) - y_0 - y_1 z^{-1}] - 4z[Y(z) - y_0] + Y(z) = R(z)$$

from which

$$(4z^2 - 4z + 1)Y(z) = 4z^2 + (12 - 4)z + R(z)$$

Dividing by 4, and rearranging

$$Y(z) = \frac{1}{(z - \frac{1}{2})^2} [z^2 + 2z + \tfrac{1}{4}R(z)]$$

The characteristic polynomial has a repeated root $z = \frac{1}{2}$, and is a perfect square. It has been noted previously that if the result

$$\mathcal{Z}\{K^n\} = 1/(1 - Kz^{-1})$$

is differentiated with respect to K, we obtain the further result

$$\mathcal{Z}\{nK^{n-1}\} = z^{-1}/(1 - Kz^{-1})^2$$

Here we have

$$\frac{z}{p(z)} = \frac{z^{-1}}{(1 - \frac{1}{2}z^{-1})^2}$$

and so identify the original sequence as

$$\tilde{y}_n = n(\tfrac{1}{2})^{n-1} \qquad \tilde{y}_0 = 0$$

Inversion of the terms $z^2/p(z)$ and $1/p(z)$ gives $\tilde{y}_{n+1}$ and $\tilde{y}_{n-1}$. If we invert the term $R(z)/p(z)$ using the convolution theorem then, for $n \geqslant 2$,

$$\mathcal{Z}^{-1}\left\{\frac{1}{p(z)} \cdot R(z)\right\} = \sum_{m=2}^{n} \tilde{y}_{m-1} r_{n-m}$$

In this case there are non-zero factors

$$r_0 = 4 = r_{n-m} \qquad (m = n)$$
$$r_1 = 4 = r_{n-m} \qquad (m = n-1)$$

from which

$$\mathcal{Z}^{-1}\left\{\frac{1}{p(z)} \cdot R(z)\right\} = 4(\tilde{y}_{n-1} + \tilde{y}_{n-2}) \qquad n \geqslant 2$$

The solution of the difference equation can then be written

$$y_n = \tilde{y}_{n+1} + 2\tilde{y}_n + H(n-2)(\tilde{y}_{n-1} + \tilde{y}_{n-2})$$

Explicitly, this gives

$$y_n = (n+1)(\tfrac{1}{2})^n + 2n(\tfrac{1}{2})^{n-1} + H(n-2)[(n-1)(\tfrac{1}{2})^{n-2} + (n-2)(\tfrac{1}{2})^{n-3}]$$

Specific results are

$$y_0 = 1$$
$$y_1 = 2(\tfrac{1}{2}) + 2(1) = 3$$

(confirming the initial conditions), and

$$y_2 = 3(\tfrac{1}{2})^2 + 4(\tfrac{1}{2}) + 1 = \tfrac{15}{4}$$

(which is directly verifiable from the original difference equation.)

If $n \geqslant 2$ the result can be simplified on setting $H(n-2) = 1$ and extracting a factor $(\tfrac{1}{2})^{n-3}$. This leads to the equation

$$y_n = \tfrac{1}{8}(\tfrac{1}{2})^{n-3}(17n - 19) \qquad n \geqslant 2 \qquad \bullet$$

4.10 DIFFERENCE EQUATIONS OF ORDER $N > 2$

Given a difference equation of order N, it is assumed that N pieces of information are available; for example, the values of $y_0, y_1, \ldots, y_{N-1}$ might be specified (or we might have information about y_n as $n \to \infty$, as in Worked Example 4.1). Otherwise, the procedure following transformation is a generalization of the method (or methods) available for the cases already discussed. Unless long division is invoked, the only difference is that we now identify the original sequence $\tilde{y}_n$ by considering $\mathscr{Z}^{-1}\{z^{N-1}/p_N(z)\}$, before applying delay theorems.

If the characteristic polynomial $p_N(z)$ has repeated roots, we can extend the concept of differentiation of the transform of K^n, discussed in the previous section, to achieve results in which the denominator in the transform is $(1 - Kz^{-1})^M$, where $M > 2$. In the same way, should $p_N(z)$ have a repeated quadratic factor we can consider the results of differentiating the transform of an appropriate hyperbolic or trigonometric sequence.

Worked Example 4.6

Given the third-order difference equation

$$y_{n+3} - 3y_{n+2} + 3y_{n+1} - y_n = 0$$

and initial values $(y_0, y_1, y_2) = (\dot{1}, 0, 1)$, find the sequence $\{y_n\}$.

Solution

We transform the equation using delay property Z1(ii) of Table 3.1, now using it with $m = 1, 2$ and 3, as this is a third-order forward difference equation, and obtain

$$z^3[Y(z) - y_0 - y_1 z^{-1} - y_2 z^{-2}] - 3z^2[Y(z) - y_0 - y_1 z^{-1}]$$
$$+ 3z[Y(z) - y_0] - Y(z) = 0$$

Incorporating the initial data and rearranging gives $p(z) = (z-1)^3$ and

$$Y(z) = \frac{1}{(z-1)^3}[z^3 - 3z^2 + 4z]$$

$$= \frac{1}{(1-z^{-1})^3} - \frac{3z^{-1}}{(1-z^{-1})^3} + \frac{4z^{-2}}{(1-z^{-1})^3} \qquad (4.16)$$

after division of numerator and denominator terms by z^3. If we differentiate the result obtained earlier, namely

$$\mathcal{Z}\{nK^{n-1}\} = \frac{z^{-1}}{(1-Kz^{-1})^2}$$

with respect to K, we obtain the further transform pair

$$\mathcal{Z}\{n(n-1)K^{n-2}\} = \frac{2z^{-2}}{(1-Kz^{-1})^3} \qquad (4.17)$$

(This corresponds to the term involving

$$\frac{z^{N-1}}{p_N(z)} = \frac{z^2}{(z-1)^3} = \frac{z^{-2}}{(1-z^{-1})^3}$$

and $K = 1$ is to be selected.)

Inverting (4.16), we first use the result (4.17) to identify the sequence

$$\tilde{y}_n = \mathcal{Z}^{-1}\left\{\frac{z^{-2}}{(1-z^{-1})^3}\right\} = \frac{1}{2}n(n-1)$$

Next we use the delay property twice, noting that $\tilde{y}_0 = \tilde{y}_1 = 0$, to obtain

$$\mathcal{Z}^{-1}\left\{\frac{z \cdot z^{-2}}{(1-z^{-1})^3}\right\} = \tilde{y}_{n+1} = \frac{1}{2}(n+1)n$$

and

$$\mathscr{L}^{-1}\left\{\frac{z^2 \cdot z^{-2}}{(1-z^{-1})^3}\right\} = \tilde{y}_{n+2} = \frac{1}{2}(n+2)(n+1)$$

From (4.16) we then have the solution

$$y_n = \tfrac{1}{2}[(n+1)(n+1) - 3(n+1)n + 4n(n-1)]$$

which simplifies to $y_n = (n-1)^2$. ●

This example is quite straightforward. More generally, some partial fraction analysis will be needed, depending on the nature of the factors of $p(z)$.

4.11 A BACKWARD DIFFERENCE EQUATION

The procedure followed is much the same as that applied to a forward difference equation, except that we shall make use of delay property Z1(i) of Table 3.1 when transforming the equation. Further, initial conditions might be written as specified values of y_{-1}, y_{-2}. (This is because approximation of a derivative at $n = 0$ will include such numbers if the approximation is based on backward differences, although we still assume that $y_n = 0$ on an infinite half-range, e.g. $n < -2$.)

A second-order equation might take the form

$$y_n + c_1 y_{n-1} + c_2 y_{n-2} = r_n \tag{4.18}$$

for example, and the solution required is the sequence $\{y_n\}$ for $n \geqslant 0$ given prescribed values of y_{-1} and y_{-2}.

The delay result obtained in Section 3.11 can be rewritten for present purposes as

$$\mathscr{L}\{y_{n-m}\} = z^{-m}\left[Y(z) + \sum_{k=1}^{m} y(-k)z^k\right]$$

whence (4.18), after transformation, becomes

$$Y(z) + c_1 z^{-1}[Y(z) + y_{-1}z] + c_2 z^{-2}[Y(z) + y_{-1}z + y_{-2}z^2] = R(z)$$

from which

$$Y(z) = \frac{1}{1 + c_1 z^{-1} + c_2 z^{-2}}[-(c_1 y_{-1} + c_2 y_{-2}) - c_2 y_{-1}z^{-1} + R(z)]$$

$$\tag{4.19}$$

This should be compared with (4.6) describing the result of transforming a second-order *forward* difference equation. In the latter case, it is necessary to convert from positive to negative powers of z at some point, if using transforms written in terms of z^{-1}. We observe in (4.19) that $Y(z)$ is already expressed in terms of z^{-1}. Otherwise inversion of the constituent terms is effected as described in previous sections, in the context of forward difference equations.

Worked Example 4.7

Solve the first-order difference equation

$$y_n + 2y_{n-1} = r_n \qquad y_{-1} = 4$$

given that

$$r_n = \begin{cases} 6 & n \geqslant 0 \\ 0 & n < 0 \end{cases}$$

Solution

Transforming the equation and noting that the transform ZT2 of Table 3.2 can be applied to $\{r_n\}$, we have

$$Y(z) + 2z^{-1}[Y(z) + y_{-1}z] = \frac{6}{1 - z^{-1}}$$

from which

$$Y(z) = \frac{1}{1 + 2z^{-1}}\left[\frac{6}{1 - z^{-1}} - 8\right]$$

Expressing the first term in partial fraction form,

$$Y(z) = \frac{6}{2 - (-1)}\left(\frac{2}{1 + 2z^{-1}} + \frac{1}{1 - z^{-1}}\right) - \frac{8}{1 + 2z^{-1}}$$

$$= -\frac{4}{1 + 2z^{-1}} + \frac{2}{1 - z^{-1}}$$

of which the inverse is

$$y_n = -4(-2)^n + 2(1)^n = 2[1 + (-2)^{n+1}]$$

(We note that this gives $y_{-1} = 4$, as required. Note also that we chose to make use of the explicit transform of $R(z)$, $\{r_n\}$ being a non-terminating sequence.) ●

In the next section, we shall be concluding the discussion of difference equations by comparing the solutions of a particular problem described alternatively by a forward difference equation and by a backward difference equation.

It is appropriate to say at this stage, however, that there is no reason why a difference equation should belong to either category: it might be presented in a form in which *both* delay properties are used in the transformation process. Also, if a difference equation is obtained by approximating a differential equation (describing the response of an electrical network or some other system) then, of course, the sampling interval $\Delta t = T$ appears in the approximation of any derivative. If it is wished to obtain a solution without first specifying what is the value of T (for example, with a view to subsequently examining the effect of changing T), then T will appear as a parameter in the difference equation, and properties and pairs of the Z-transform as written in general terms in Tables 3.1 and 3.2 are used, so that the solution appears in the form $y_n = y(nT)$. If the sampling interval is preset, then the value of T is subsumed in the coefficients (e.g. c_0 and c_1) appearing in the difference equation. It should be taken that, in the worked examples of this chapter, a preset value of T is implied. (In Problem 4.9, the reader is asked first to obtain a solution in terms of an unspecified value of T.)

4.12 A SECOND-ORDER EQUATION: COMPARISON OF THE TWO METHODS

When comparing a backward difference model with the forward difference equivalent, we can regard either as a shifted version of the other. The solutions generate the same sequence values, but from a different origin on the axis.

Worked Example 4.8

Let us consider a system whose forward difference description is provided by the equation

$$y_{n+2} + 2y_{n+1} - 3y_n = 0 \qquad y_0 = 1, y_1 = 0 \qquad (4.20)$$

We postulate that the backward difference problem

$$y_k + 2y_{k-1} - 3y_{k-2} = 0 \qquad y_{-2} = 1, y_{-1} = 0 \qquad (4.21)$$

is equivalent. (We have changed the suffix to k in (4.21) to facilitate subsequent comparison of $\{y_n\}$ and $\{y_k\}$.)

Solution

We will use notation $Y^+(z) = \mathscr{Z}\{y_n\}$ and $Y^-(z) = \mathscr{Z}\{y_k\}$ to distinguish between the forward and backward descriptions, and similarly use the notation y_n^+ and y_k^- for the respective solutions.

From (4.20), we obtain the transformed equation

$$z^2[Y^+(z) - y_0 - y_1 z^{-1}] + 2z[Y^+(z) - y_1] - 3Y^+(z) = 0$$

giving

$$(z^2 + 2z - 3)Y^+(z) = z^2 + 2z$$

The characteristic polynomial has real factors. After division of both sides by z^2 and obtaining partial fractions, we have

$$Y^+(z) = \frac{1}{4}\left(\frac{1}{1 + 3z^{-1}} + \frac{3}{1 - z^{-1}}\right)$$

from which

$$y_n^+ = \tfrac{1}{4}[(-3)^n + 3(1^n)] = \tfrac{3}{4}[1 - (-3)^{n-1}] \tag{4.22}$$

for $n = 0, 1, 2, \ldots$, and the initial conditions are met.

Transforming (4.21) gives

$$Y^-(z) + 2z^{-1}[Y^-(z) + y_{-1}z] - 3z^{-2}[Y^-(z) + y_{-1}z + y_{-2}z^2] = 0$$

Substituting initial values and rearranging slightly leads to the equation

$$(1 + 2z^{-1} - 3z^{-2})Y^-(z) = 3$$

From this we obtain $Y^-(z)$ in the partial fraction form

$$Y^-(z) = \frac{3}{4}\left(\frac{3}{1 + 3z^{-1}} + \frac{1}{1 - z^{-1}}\right)$$

which inverts to give

$$y_k^- = \tfrac{3}{4}[3(-3)^k + (1)^k] = \tfrac{3}{4}[1 - (-3)^{k+1}] \tag{4.23}$$

for $k = -2, -1, 0, 1, \ldots$, which we note agrees with the initial conditions $y_{-2} = 1$, $y_{-1} = 0$. If we shift y_k^- through two units to the right by putting $k = n - 2$, then the right-hand side of (4.23) becomes

$$\tfrac{3}{4}[1 - (-3)^{n-1}] \qquad \text{for } n = 0, 1, 2, \ldots$$

which is the result given in (4.22). The two approaches therefore *do* generate the same sequence, and only differ in the integer suffix assigned to the first term in the sequence (i.e. $n = 0$ or $k = -2$). ●

SUMMARY

In Chapter 4 we have considered how linear difference equations with constant coefficients might arise from the approximation of a differential equation or as the immediate description of a physical system. We have instanced (in Section 4.4) a possible approach to the solution of a difference equation without recourse to transformation methods, but for the most part have discussed formal methods based on the properties of the Z-transform which were developed in Chapter 3.

The analogy between techniques applicable to differential equations and difference equations has emerged in several contexts. In particular, we have illustrated, in the case of an inhomogeneous difference equation, that either a convolution method or direct use of the transform $R(z)$ of an input sequence $\{r_n\}$ can be considered during the inversion process.

PROBLEMS

4.1 In Worked Example 4.2, transform methods were illustrated by deriving the solution of the problem

$$y_{n+2} - 4y_{n+1} + 3y_n = 0 \qquad y_0 \text{ and } y_1 \text{ arbitrary}$$

in the form

$$y_n = \left(\frac{3}{2}y_0 - \frac{1}{2}y_1\right) + \left(-\frac{y_0}{2} + \frac{y_1}{2}\right)3^n$$

Obtain this result by assuming *ab initio* that there are solutions of the form $y_n = BK^n$.

4.2 In Worked Example 4.3, the solution to the problem

$$y_{n+2} - 4y_{n+1} + 5y_n = 0 \qquad y_0 = \tfrac{1}{2}, y_1 = 3$$

was shown to be

$$y_n = 5^{n/2}[2\sin(na) + \tfrac{1}{2}\cos(na)]$$

in which $a = \tan^{-1}\tfrac{1}{2}$. Show that this can also be obtained by looking for solutions of the form $y_n = BK^n$.

4.3 In Worked Example 4.6 we showed that the problem

$$y_{n+3} - 3y_{n+2} + 3y_{n+1} - y_n = 0 \qquad y_0 = 1, y_1 = 0, y_2 = 1$$

has the solution $y_n = (n-1)^2$. Show that substitution of

$y_n = BK^n$ into the equation identifies only the sequence $K^n = 1^n$, and so does not provide the general solution of the difference equation (of order 3). Assuming instead trial solutions of the form $y_n = B(n)K^n$, where K has been found equal to 1, show that if $B(n) = b_0 + b_1 n + b_2 n^2$ we can satisfy the difference equation identically and obtain the solution previously obtained by transform methods. (Note that this is equivalent to the method of variation of parameters used when the assumption that a homogeneous differential equation has solutions $y = B e^{\lambda t}$ leads to an auxiliary – or characteristic – equation in λ having equal roots. In association with any repeated root, say λ_i, we assume instead solutions $y = B(t) e^{\lambda_i t}$, and find that $B(t)$ is a polynomial.)

4.4   Use transform methods to show that the solution of the homogeneous difference equation

$$3y_{n+2} - 5y_{n+1} + 2y_n = 0 \qquad y_0 = 1, y_1 = 0$$

is

$$y_n = 2\left[\left(\frac{2}{3}\right)^{n-1} - 1 \right]$$

4.5   Use transform methods to show that if

$$y_{n+2} - y_{n+1} + y_n = 0 \qquad y_0 = 1, y_1 = 2$$

then

$$y_n = 2\cos\left[\frac{(n-1)\pi}{3} \right]$$

4.6   (Refer to Problem 4.4).
Given a sequence $\{r_n\}$ such that $r_1 = 1, r_2 = 3$ and $r_n = 0$ if $n \neq 1, 2$, show that the inhomogeneous difference equation

$$3y_{n+2} - 5y_{n+1} + 2y_n = r_n \qquad y_0 = 1, y_1 = 0$$

has a solution

$$y_n = 2\left[\left(\frac{2}{3}\right)^{n-1} - 1 \right] + H(n-3)\frac{1}{3}\left[12 - 11\left(\frac{2}{3}\right)^{n-3} \right]$$

using convolution methods to obtain the second of these two terms.

4.7   Show that, if $Y(z)$ is the transform of the inhomogeneous equation in Problem 4.6,

$$\lim_{z \to 1} (1 - z^{-1})Y(z) = \lim_{n \to \infty} y_n = 2$$

4.8 Given that $r_0 = 1$ and $r_n = 0$ if $n \neq 0$, solve the third-order difference equation

$$y_{n+3} - 6y_{n+2} + 12y_{n+1} - 8y_n = r_n$$

with initial values $(y_0, y_1, y_2) = (1, 1, 0)$. (Do *not* use convolution in this example.) Use previously established results for $\mathscr{Z}\{nK^{n-1}\}$ and $\mathscr{Z}\{n(n-1)K^{n-2}\}$ as necessary, together with appropriate delay properties to invert $Y(z)$.

(*Hint*: Having found the original sequence $\tilde{y}_n$, you should note that $\tilde{y}_0 = \tilde{y}_1 = 0$ and use the delay property to obtain $\tilde{y}_{n+1}$ and $\tilde{y}_{n+2}$. You will also need $\tilde{y}_{n-1}$, which is non-zero for $n > 2$.)

Show that, after simplification, the solution can be written

$$y_n = (n-2)2^{n-3}[\tfrac{1}{2}(n-1)H(n-3) - 4]$$

4.9 (i) A circuit is described by the differential equation

$$\frac{dv}{dt} + 2v = i(t)$$

in which the input is a unit current $i(t) = 1$ and the output voltage is initially $v(0) = 2$. Use Laplace transform methods to show that the solution of the continuous time problem is

$$v(t) = \tfrac{1}{2}(1 + 3e^{-2t})$$

(ii) If the derivative term is approximated by $\Delta v/\Delta t$, using a backward difference and sampling time T, show that the system described in (i) is modelled by the difference equation

$$(1 + 2T)v_n - v_{n-1} = Ti_n$$

in which $v_n = v(nT)$ and $i_n = i(nT) = 1$, and $v_{-1} = 2$. Use Z-transform methods to show that the voltage sequence is

$$v(nT) = \frac{1}{2}\left[1 + 3\left(\frac{1}{1 + 2T}\right)^{n+1}\right]$$

(iii) Show that at $t = 2$ the continuous solution gives $v(2) = 0.527$, and that if $T = 1$ the discrete solution gives $v(2) = 0.555$. Confirm that reducing T leads to a better result.

4.10 Given the difference equation

$$y_{n+1} + y_n = \sqrt{2}\sin\left(\frac{n\pi}{4}\right) \qquad y_0 = 0$$

use Z-transform methods, partial fractions and tables to show
that

$$y_n = \left(1 - \frac{1}{\sqrt{2}}\right)(-1)^n - \cos\left(\frac{n\pi}{4}\right) + \sin\left[\frac{(n+1)\pi}{4}\right]$$

4.11 From the Z-transform pairs ZT6 and ZT8 of Table 3.2,
differentiate to obtain $\mathscr{Z}\{n\cos(an)\}$ and $\mathscr{Z}\{n\sin(an)\}$. Deduce
that

$$\mathscr{Z}\left\{n\cos\left(\frac{n\pi}{2}\right)\right\} = -\frac{2z^{-2}}{(1+z^{-2})^2}$$

and

$$\mathscr{Z}\left\{n\sin\left(\frac{n\pi}{2}\right)\right\} = \frac{z^{-1} - z^{-3}}{(1+z^{-2})^2}$$

4.12 (i) Given

$$y_{n+2} - y_n = \sin\left(\frac{n\pi}{2}\right) \qquad y_0 = y_1 = 1$$

show that the transform $Y(z)$ has partial fractions of the form

$$Y(z) = \frac{A}{1 - z^{-1}} + \frac{B}{1 + z^{-1}} + \frac{Cz^{-1} + D}{1 + z^{-2}}$$

Find the constants A, B, C and D and hence the result

$$y_n = \frac{1}{4}\left[5 - (-1)^n - 2\sin\left(\frac{n\pi}{2}\right)\right]$$

Show that $\{y_n\}$ is a sequence which repeats the four values
$\{1, 1, 1, 2\}$.

(ii) Find the transform $Y(z)$ given the equation

$$y_{n+2} + y_n = \sin\left(\frac{n\pi}{2}\right) \qquad y_0 = y_1 = 1$$

Invert this using the results of Problem 4.11 and delay properties as appropriate to show that, in this case, the solution
is the unbounded sequence

$$y_n = \cos\left(\frac{n\pi}{2}\right) + \sin\left(\frac{n\pi}{2}\right) - \frac{1}{2}(n-1)\cos\left[\frac{(n-1)\pi}{4}\right]$$

and values of four successive terms are given by $\{1, 1 - 2p, -1, 2p\}$ on substituting $n = 4p, 4p + 1, 4p + 2, 4p + 3$.

5

The discrete Fourier transform

In Chapter 2 we developed properties of the (continuous-time) direct Fourier transform and the inverse Fourier transform, the two constituting an integral pair. Whereas Fourier *series* analysis is largely concerned with functions which are treated as being periodic, the Fourier *transform* provides an instrument for the analysis of non-periodic functions. Applications range from the processing of signals in many contexts (engineering, cardiology, speech wave-forms, etc.), to the analysis of power spectra, as well as the solution of differential equations.

In Chapter 3 we considered in some detail the conversion of a continuous-time signal to discrete form, and also the effect of time-domain sampling on the frequency-domain spectrum (or transform). At that stage we did not refer to any particular transform (discrete or not). It is the purpose of this chapter to consider how the Fourier transform pair of infinite integrals can be replaced by finite expressions amenable to digital computation. (Obviously it will be necessary to confirm that such expressions *do* constitute an inversion pair.) We shall then develop some properties of the discrete transform, analogous to those obtained for the integral transform.

In Chapters 1 and 2 we referred to the relationship between the one-sided Laplace and Fourier transforms. Similarly, if the Z-transform is the Laplace transform of a sampled signal, we can expect there to be a relationship between that and the Fourier transform of a sampled signal. For present purposes, any comparison ends at that point. At the end of Section 3.6, in which we examined sampling in the time domain, we referred (without detail) to sampling in the frequency domain. In Section 3.9, we defined the Z-transform in terms of a continuous variable z, and in (3.14) gave the inverse of a transform $X(z)$ in the form of an integral. In order to invert a Z-transform, however, we did not use that result, and neither did we use what might be called 'sampled values' of $X(z)$. (Time t was replaced by discrete values nT, but there was no corresponding modification of the variable z.) Instead we used transform properties and other techniques, including in particular tabulated pairs. In this chapter we shall proceed immediately to produce expressions in which both the transform and the inverse are expressed in terms of discrete variables, in time *and* in frequency.) Analysis will then involve two sets of discrete data, and although we will consider *properties* of the discrete Fourier transform (DFT), we shall not need any equivalent of the table of Z-transform pairs (Table 3.2).

We shall also return to the matter of convolution (always associated with the identification of a transfer function) and in Section 5.6 will introduce the related concept of correlation, in the same context.

5.1 APPROXIMATING THE EXPONENTIAL FOURIER SERIES

So that we can consider values of a function on a finite range, we take as our starting point the series representation of a function $x(t)$. For example, if we return to Chapter 1 and put

$$x(t) = \sum_{n=-\infty}^{\infty} X_n e^{j2\pi n f_0 t} \tag{5.1}$$

then

$$X_n = \frac{1}{T_0} \int_{T_0} x(t) e^{-j2\pi n f_0 t} dt \tag{5.2}$$

in which $f_0 = 1/T_0$. The implication is that $x(t) = x(t + T_0)$, whether $x(t)$ is actually periodic or not. The limits of integration in (5.2) can be any values $[a, b]$ such that $b - a = T_0$.

As stated at the end of Section 1.2, such expressions can be regarded as constituting a transform pair. We note that the sum in (5.1) is a sum over the discrete integer variable n, but is infinite. In (5.2), the integration variable t is continuous, but the range of integration, say $[0, T_0]$, is finite.

Let us replace the integral in (5.2) by a rectangular approximation based on left-hand ordinate values, as shown in Fig. 5.1. If there are N intervals covering the range $[0, T_0]$ then sampling occurs at points

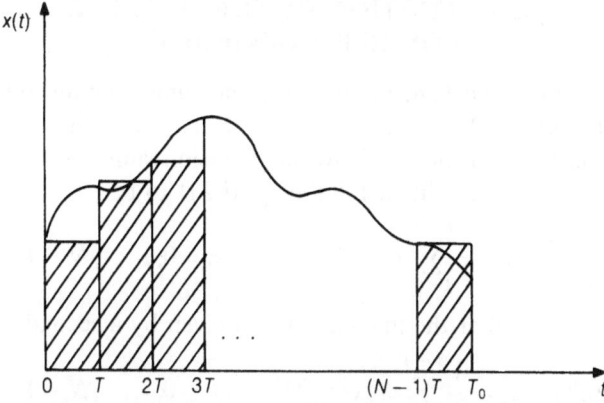

Fig. 5.1 Approximation of the coefficient integral, X_n.

kT where T is the sampling interval $\Delta t = T = T_0/N$. Summing the areas of the N rectangles and substituting in (5.2), we obtain

$$X_n = \frac{1}{T_0}\left\{\sum_{k=0}^{N-1} x_s(kT)\,e^{-j2\pi nf_0kT} \cdot \frac{T_0}{N}\right\}$$

Since $f_0 T = (1/T_0)\cdot(T_0/N)$, this simplifies to give

$$X_n = \frac{1}{N}\sum_{k=0}^{N-1} x_s(kT)\,e^{-j(2\pi/N)nk} \qquad n = 0, 1, 2, \ldots, N-1 \qquad (5.3)$$

(Observe that the N sampled values in this expression do not include $x(T_0)$.)

The sum in (5.3) is a sum over N values $k = 0, 1, 2, \ldots, N-1$, and the calculation must be repeated for each value of n. We next approximate (5.1) by truncation, so that that sum also is a sum of N terms, and we contemplate time-series values

$$x_k = \sum_{n=0}^{N-1} X_n\,e^{j(2\pi/N)nk} \qquad (5.4)$$

in which we have replaced t in (5.1) by kT_0/N, noting that $f_0 = 1/T_0$, and have put $x_s(kT) = x_k$. Whether we can still claim to have a transform pair after these approximations will be considered in the next section.

5.2 DEFINITION OF THE DISCRETE FOURIER TRANSFORM

Equations (5.3) and (5.4) in effect are the sums defining the direct DFT and, as will be seen, the inverse DFT, except that it is usual to transfer the multiplier $1/N$ to the inversion sum.

We will define the direct DFT of $\{x_s(kT)\}$ to be

$$X\left(\frac{n}{NT}\right) = X_n = \sum_{k=0}^{N-1} x_k e^{-j(2\pi/N)nk} \qquad n = 0, 1, \ldots, N-1 \qquad (5.5)$$

It is postulated that the inverse DFT is given by a second sum,

$$x_s(kT) = x_k = \frac{1}{N} \sum_{n=0}^{N-1} X_n e^{j(2\pi/N)nk} \qquad k = 0, 1, \ldots, N-1 \qquad (5.6)$$

Now we can address the question of whether these equations provide us with a transform pair. We need to show that if we use (5.5) to substitute for X_n in (5.6) then the resulting double sum reduces to x_k. If so, then (5.6) *does* invert the transform defined by (5.5).

As written, the sum in (5.6) is a summation over varying n for some fixed value of k, whereas in (5.5) k is varying. To avoid confusion, we rewrite (5.5) in terms of some other summation integer before substitution. For example, on replacing k by m,

$$X_n = \sum_{m=0}^{N-1} x_m e^{-j(2\pi/N)nm} \qquad n = 0, 1, \ldots, N-1$$

Putting this in the right-hand side of (5.6) gives the expression

$$EX = \frac{1}{N} \sum_{n=0}^{N-1} \left\{ \sum_{m=0}^{N-1} x_m e^{-j(2\pi/N)nm} \right\} e^{j(2\pi/N)nk}$$

in which brackets imply that the summation over m takes place first. (We have introduced EX so as not to prejudge the issue.) If this is rewritten so that summation over n takes place first, then, since x_m is the only quantity independent of n, we obtain

$$EX = \frac{1}{N} \sum_{m=0}^{N-1} \left\{ \sum_{n=0}^{N-1} e^{j(2\pi/N)n(k-m)} \right\} x_m \qquad (5.7)$$

Consider the sum in brackets. Integer k is some fixed number in the range $0, 1, \ldots, N-1$. If $m \neq k$ then the sum can be recognized as a finite geometric series

$$S_N = a + ar + \cdots + ar^{N-1} = \sum_{n=0}^{N-1} ar^n$$

in which $a = 1$ and $r = e^{j(2\pi/N)(k-m)}$. The sum of a finite geometric progression is given by $S_N = a(1 - r^N)/(1 - r)$, and so the inner sum of (5.7) is

$$S_N = \frac{1 - e^{j2\pi(k-m)}}{1 - e^{j(2\pi/N)(k-m)}}$$

The exponential in the numerator of this expression is equal to 1, as the index is an integer multiple of $2\pi j$. If $m \neq k$, then in the denominator $(k - m)/N$ is not an integer. In that case, the numerator is $1 - 1 = 0$, whereas the denominator is non-zero. We conclude that if $m \neq k$, $S_N = 0$ and the coefficient of x_m is zero.

If $m = k$ then $r = 1$, and S_N comprises N terms all equal to 1, whence $S_N = N$. Equation (5.7) therefore reduces to

$$EX = \frac{1}{N} \cdot N \cdot x_k = x_k$$

so justifying the assertion that (5.6) *is* the inverse of the transform defined by (5.5).

5.3 ESTABLISHING THE INVERSE

In the previous section we confirmed, in terms of a general integer N, that (5.5) and (5.6) constitute a transform pair. Those results will now be applied to a very simple example. The intention is to show that, even in such a case, direct use of these equations for computational purposes does *not* appear to provide an efficient method of calculation.

Worked Example 5.1

Given $N = 8$ and sampled values

$$\{x_k\} = (1, 1, 1, 0, 0, 0, 0, 0) \quad \text{for} \quad k = 0, 1, \ldots, 7$$

find the DFT $\{X_n\}$, and verify that $\{x_k\}$ is recovered on inversion.

Solution

Noting that $x_0 = x_1 = x_2 = 1$ and that $x_k = 0$ for $k \geqslant 3$, from (5.5) the elements of the transform are given by computing

$$X_n = 1 + e^{-j\pi n/4} + e^{-j\pi n/2} \tag{5.8}$$

for $n = 0, 1, \ldots, 7$ in the case where $N = 8$.

If wished, this can be replaced by the equation

$$X_n = \left[1 + \cos\left(\frac{n\pi}{4}\right) + \cos\left(\frac{n\pi}{2}\right) \right] - j\left[\sin\left(\frac{n\pi}{4}\right) + \sin\left(\frac{n\pi}{2}\right) \right]$$

which can be simplified further, using double angle formulae, to read

$$X_n = \left[1 + 2\cos\left(\frac{n\pi}{4}\right) \right]\left[\cos\left(\frac{n\pi}{4}\right) - j\sin\left(\frac{n\pi}{4}\right) \right] \qquad (5.9)$$

Using either (5.8) or (5.9) we obtain, explicitly,

$$\begin{bmatrix} X_0 \\ X_1 \\ X_2 \\ X_3 \\ X_4 \\ X_5 \\ X_6 \\ X_7 \end{bmatrix} = \begin{bmatrix} 3 \\ (1 + 1/\sqrt{2}) - j(1 + 1/\sqrt{2}) \\ -j \\ (1 - 1/\sqrt{2}) + j(1 - 1/\sqrt{2}) \\ 1 \\ (1 - 1/\sqrt{2}) - j(1 - 1/\sqrt{2}) \\ +j \\ (1 + 1/\sqrt{2}) + j(1 + 1/\sqrt{2}) \end{bmatrix} \qquad (5.10)$$

With $N = 8$, the inversion sum (5.6) reads

$$x_k = \frac{1}{8} \sum_{n=0}^{7} X_n e^{j\pi nk/4} \qquad k = 0, 1, \ldots, 7 \qquad (5.11)$$

Now, whether we substitute values of X_n as given by (5.8) or (5.9), or use the explicit values listed in (5.10), there is no obvious simplification leading to a closed-form expression for x_k. (Indeed, we do not expect to be able to write x_k as a simple function of k, given the nature of the sequence defined in the question.)

We will compute the values of just two terms in the sequence, say x_2 and x_3, looking for *ad hoc* simplifications in each particular summation.

If we decide to use (5.8), then putting $k = 2$ in (5.11) gives

$$x_2 = \frac{1}{8} \sum_{n=0}^{7} \{ e^{j\pi n/2} + e^{j\pi n/4} + 1 \}$$

after multiplying X_n by $e^{j\pi n/2}$. In this case, it is worth noticing that the term in brackets is the complex conjugate of X_n as given by (5.8), whence

$$x_2 = \frac{1}{8} \sum_{n=0}^{7} \bar{X}_n$$

We can now use the explicit results given in (5.10) and sum to conclude that

$$x_2 = \frac{1}{8} \cdot 8 = 1$$

(as expected).

Proceeding in the same way, but putting $k = 3$ in (5.11), leads to the expression

$$x_3 = \frac{1}{8} \sum_{n=0}^{7} \{1 + e^{-\pi n/4} + e^{-j\pi n/2}\} e^{j\pi 3n/4}$$

There is no simplification of this comparable to that noticed when considering x_2, but rather than use explicit values of X_n it might be simpler to modify the last equation. We can write each term in the form

$$e^{j\pi n/2}[e^{j\pi n/4} + 1 + e^{-j\pi n/4}] = e^{j\pi n/2}\left[1 + 2\cos\left(\frac{n\pi}{4}\right)\right]$$

so that we need evaluate only two particularly simple functions. We then obtain

$$x_3 = \frac{1}{8}[1 \cdot (3) + j(1 + 2/\sqrt{2}) - 1 \cdot (1) - j(1 - 2/\sqrt{2}) + 1 \cdot (-1)$$
$$+ j(1 - 2/\sqrt{2}) - 1 \cdot (1) - j(1 + 2/\sqrt{2})] = 0 \qquad \bullet$$

If, in Worked Example 5.1, it appeared that the computational effort expended seemed disproportionate (given the simple nature of the problem), the intention was to convey exactly that impression. In this example we took $N = 8$, whereas in a more realistic situation the number of samples in a sequence being analysed is likely to be in the range $1000 < N < 10\,000$. Furthermore, samples $\{x_k\}$ might be complex or multidimensional numbers (i.e. vectors). Even with access to considerable computing power, one should not base spectral analysis on the direct use of the transform pair as given by (5.5) and (5.6).

In Chapter 6 and (in particular) Chapter 7 efficient methods of computing the DFT and its inverse will be discussed in some detail. Without anticipating the development of such methods, we hope that some clues as to where we might look for economies have emerged from Worked Example 5.1.

First, in both the DFT and the inverse calculations we are concerned with cyclic (i.e. exponential or trigonometric) functions. Second, while calculating the spectrum $\{X_n\}$, it appears that various sub-calculations are being repeated. If some method of breaking down the X_n-summation can be found, we ought to be able to devise a method of nested calculation whereby any interim result is obtained once and once only, and is then incorporated where it appears in the computation of each element $X_n, n = 0, 1, \ldots, N - 1$.

That the direct use of (5.5) *does* involve repeated calculations can be demonstrated by means of an example in which the input sequence $\{x_k\}$ is not specified. We conclude this section with a simple example, the case $N = 4$.

Worked Example 5.2

Obtain the four-point DFT of time-domain samples $\{x_0, x_1, x_2, x_3\}$ from the defining sum, simplifying the result as far as possible.

Solution

From (5.5), with $N = 4$, the spectrum is given by

$$X_n = \sum_{k=0}^{3} x_k e^{-j\pi nk/2} \qquad n = 0, 1, 2, 3$$

For convenience, put $W = W_4 = e^{-j\pi/2}$, so that

$$X_n = \sum_{k=0}^{3} x_k W^{nk} \qquad n = 0, 1, 2, 3$$

Explicitly, this gives expressions

$$X_0 = x_0 + x_1 + x_2 + x_3 \qquad \text{(because } W^0 = 1)$$
$$X_1 = x_0 + x_1 W^1 + x_2 W^2 + x_3 W^3$$
$$X_2 = x_0 + x_1 W^2 + x_2 W^4 + x_3 W^6$$
$$X_3 = x_0 + x_1 W^3 + x_2 W^6 + x_3 W^9$$

However, the coefficient matrix (whose elements are powers of W) can be simplified considerably if we refer to the Argand diagram of $\{W^p\}$ shown in Fig. 5.2. Since $W_4^2 = -1, W_4^4 = W_4^0 = +1, W^1 = -j$ and $W_4^3 = -W_4^1 = +j$, the previous four equations can be rewritten

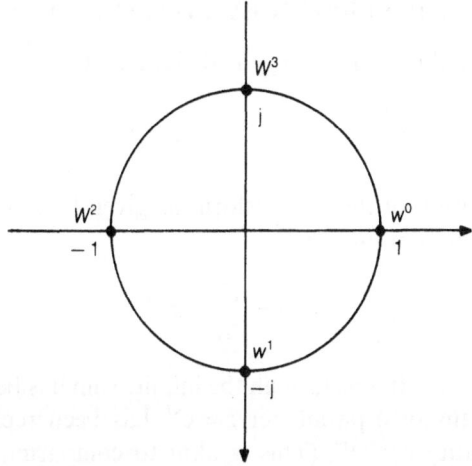

Fig. 5.2 Powers of $W = e^{-j\pi/2}$.

in the form

$$X_0 = (x_0 + x_2) + (x_1 + x_3)$$
$$X_1 = (x_0 - x_2) - j(x_1 - x_3)$$
$$X_2 = (x_0 + x_2) - (x_1 + x_3)$$
$$X_3 = (x_0 - x_2) + j(x_1 - x_3)$$

Two further points should be noted. First, each bracketed term on the right-hand side appears twice. (It need be calculated once only.) Second, these terms are combinations of two values of x_k, the suffixes being both even or both odd. In fact if $\mathbf{x}_k$ is put into bit-reversed order (see Section 3.1 and Problem 3.1), we obtain

$$\mathbf{x}_{br} = E\mathbf{x}_{nat} = \{x_0, x_2, x_1, x_3\}^T$$

in which the permutation matrix is

$$E = \begin{bmatrix} 1 & 0 & 0 & 0 \\ 0 & 0 & 1 & 0 \\ 0 & 1 & 0 & 0 \\ 0 & 0 & 0 & 1 \end{bmatrix}$$

The terms in $\{X_n\}$ are then combinations of pairs of consecutive inputs after this reordering. ●

Such considerations form the basis of the methods to be developed in the next two chapters.

5.4 INVERSION BY CONJUGATION

If we compare the DFT of $\{x_k\}$ as defined by (5.5)

$$X_n = \sum_{k=0}^{N-1} x_k e^{-j(2\pi/N)nk} \qquad n = 0, 1, \ldots, N-1$$

and the definition of the Z-transform as given by (3.13), written in terms of summation integer k,

$$X(z) = \sum_{k=0}^{\infty} x_k z^{-k}$$

we see an analogy. It is as though the infinite sum has been truncated and the (continuous) parameter $z = e^{sT}$ has been replaced by the discrete quantity $e^{j(2\pi/N)n}$. (This is akin to comparing the Laplace transform and the continuous Fourier transform.)

However, as stated in the introduction to this chapter, we shall not make any use of the relationship between the DFT and the Z-transform – not only because we now have discrete data in both time and frequency domains. The DFT sum and the inversion sum exhibit the same type of symmetry as do the integral Fourier transform pair, and we can exploit that symmetry by taking the complex conjugates of (5.5) and (5.6).

First, let us take the complex conjugate of (5.5) and, at the same time, divide by N. We obtain

$$\frac{\bar{X}_n}{N} = \frac{1}{N} \sum_{k=0}^{N-1} \bar{x}_k e^{j(2\pi/N)nk} \qquad n = 0, 1, \ldots, N-1 \qquad (5.12)$$

This has the same structure as has the inversion sum given in (5.6),

$$x_k = \frac{1}{N} \sum_{n=0}^{N-1} X_n e^{j(2\pi/N)nk} \qquad k = 0, 1, \ldots, N-1$$

In other words, had we developed an algorithm to invert a spectrum $\{X_n\}$ then the same algorithm could be used to obtain the transform itself. Instead of X_n in the previous summation, we input values of the conjugate of the sequence $\{x_k\}$. According to (5.12), the output is then the conjugate of the spectrum $\{X_n\}$, divided by N. Since a second conjugation gives $\bar{\bar{X}}_n = X_n$, all that remains necessary is to conjugate the results given by (5.12) and multiply by N, to obtain $\{X_n\}$.

Conversely, it can be shown that the equation defining the DFT can be used both to find the transform of a known time-domain signal and to recover a signal from a known spectrum.

If we conjugate the inversion sum, (5.6), and multiply by N, the result is the expression

$$N\bar{x}_k = \sum_{n=0}^{N-1} \bar{X}_n e^{-j(2\pi/N)nk} \qquad k = 0, 1, \ldots, N-1 \qquad (5.13)$$

which can be compared with (5.5). The right-hand side of (5.13) has the same structure, and can be interpreted as saying that the DFT of the input sequence $\{\bar{X}_n\}$ is the spectrum $\{N\bar{x}_k\}$. So, to recover $\{x_k\}$ from a known spectrum $\{X_n\}$, we use the transform algorithm again, with appropriate conjugation, and finally divide by N.

Equations (5.5) and (5.13) therefore replace the transform pair as originally given by (5.5) and (5.6). (It is more usual to exploit the symmetry of the transform by using (5.5) and (5.13) then (5.6) and (5.12).) We emphasize that the coefficient matrix (of exponential multipliers) is the same in both equations and that, in consequence, any algorithm developed for the calculation of a DFT can also be used to invert a spectrum, with no modification other than conjugation of inputs and outputs and provision for the factor N.

Worked Example 5.3

Find the four-point DFT of the sequence $\{x_k\} = (0, j, 0, -j)^T, k = 0, 1, 2, 3$, and invert the resulting transform $\{X_n\}$ using the same summation method, with appropriate conjugation.

Solution

Putting $N = 4$ into (5.5) and (5.13) gives us

$$X_n = \sum_{k=0}^{3} x_k e^{-j(\pi/2)nk} \qquad n = 0, 1, 2, 3 \qquad (5.14)$$

and

$$4\bar{x}_k = \sum_{n=0}^{3} \bar{X}_n e^{-j(\pi/2)nk} \qquad k = 0, 1, 2, 3 \qquad (5.15)$$

From (5.14), substituting for $\{x_k\}$ gives

$$X_n = j[e^{-j(\pi/2)n} - e^{-j(3\pi/2)n}] \qquad n = 0, 1, 2, 3$$

which can be written, if wished, in the form

$$X_n = je^{-j\pi n}(e^{j\pi n/2} - e^{-j\pi n/2}) = -2e^{-j\pi n}\sin\left(\frac{n\pi}{2}\right)$$

Explicit results are

$$\begin{bmatrix} X_0 \\ X_1 \\ X_2 \\ X_3 \end{bmatrix} = \begin{bmatrix} 0 \\ 2 \\ 0 \\ -2 \end{bmatrix}$$

and we note that (in *this* example) $\bar{X}_n = X_n$.

From (5.15), we now have

$$4\bar{x}_k = 2[e^{-j(\pi/2)k} - e^{-j(3\pi/2)k}] \qquad k = 0, 1, 2, 3$$

$$= 4je^{-j\pi k}\sin\left(\frac{k\pi}{2}\right)$$

from which

$$4\bar{x}_k = (0, -4j, 0, 4j)^T$$

On division by 4 and taking complex conjugates, we obtain the expected result

$$\mathbf{x}_k = (0, j, 0, -j)^T \qquad\qquad \bullet$$

Although we have not yet addressed in detail the question as to how we might compute a DFT in an efficient manner, we have at least established that matters are improved when computation of an inverse requires no additional procedure.

The matter of symmetry (or duality) was germane in Chapter 2, particularly in the context of developing a table of transform pairs (Table 2.2). It has been explained that in the context of the discrete Fourier transform we are more interested in the actual processing of discrete data sets and the concept of a table of transform pairs is not relevant. On the other hand, just as the integral Fourier transform is central to the general development of other integral transforms – so is the discrete Fourier transform. For example, in Chapters 1 and 2, we related the integral Fourier transform to the Fourier sine and cosine transform and to the Laplace transforms, and considered the transforms of odd and even functions. Similarly, we can consider what discrete transforms are related to the DFT.

(Such a near-relation of the DFT, the Hartley transform, is described in Appendix B.)

For this and other reasons it is, therefore, necessary to develop properties of the DFT which are the counterparts of properties of the integral Fourier transform.

5.5 PROPERTIES OF THE DISCRETE FOURIER TRANSFORM

Selected properties of the integral Fourier transform appeared in Table 2.1. In Table 5.1 we provide some properties of the DFT, and although no attempt is made to achieve a one-to-one match between the two tables (in that properties are included in each that are not given in the other), parallels between the continuous and discrete versions emerge.

Table 5.1 Properties of the DFT

Property	Signal	Transform (spectrum)
D1 Duality	$X(k)/N$	$x(-n)$
D2 Inversion (by conjugation)	$N\bar{x}_k = \sum\limits_{n=0}^{N-1} \bar{X}_n e^{-j(2\pi/N)nK}$	n/a
D3 Time shift	$x(k-m)$	$X_n e^{-j(2\pi/N)mn}$
D4 Frequency shift	$x_k e^{j(2\pi/N)mk}$	$X(n-m)$
D5 (i) Convolution (time)	$\sum\limits_{m=0}^{N-1} x_1(m)x_2(k-m)$	$X_1(n)X_2(n)$
(ii) Products (time)	$x_1(k)x_2(k)$	$\dfrac{1}{N}\sum\limits_{m=0}^{N-1} X_1(m)X_2(n-m)$
D6 DFT of a real, even, signal	$x(k)$	$\sum\limits_{k=0}^{N-1} x_k \cos\left[\left(\dfrac{2\pi}{N}\right)nk\right]$
D7 DFT of a real, odd, signal	$x(k)$	$-j\sum\limits_{k=0}^{N-1} x_k \sin\left[\left(\dfrac{2\pi}{N}\right)nk\right]$

Worked Example 5.4

Show that if the DFT of $\{x_k\}$ is $\{X_n\}$, then the DFT of the time-domain sequence $\{X_k/N\}$ is $\{x_{-n}\}$.

Solution

From (5.6), the inverse of the spectrum $\{X_n\}$ is given by

$$x_k = \frac{1}{N}\sum_{n=0}^{N-1} X_n e^{j(2\pi/N)nk}$$

Exchanging k and n gives

$$x_n = \sum_{k=0}^{N-1} \frac{X_k}{N} e^{j(2\pi/N)nk}$$

If n is replaced by $-n$ in this, the resulting sum is, by definition, the transform of the sequence $\{X_k/N\}$. This proves the duality property, D1 of Table 5.1. ●

Inversion by means of conjugation (property D2) was discussed in Section 5.4. Proofs of other properties are obtained by methods similar to those employed when deriving those of the continuous transform, making use wherever possible of the symmetry. Proof of the time-shift property D3 is invited in Problem 5.5.

Worked Example 5.5

Show that the DFT of the sequence

$$x_k e^{j(2\pi/N)mk}$$

in which m is a fixed integer, is $X(n-m)$.

Solution

From the definition, the required transform is given by

$$\sum_{k=0}^{N-1} x_k e^{j(2\pi/N)mk} e^{-j(2\pi/N)nk} \qquad n = 0, 1, \ldots, 7$$

$$= \sum_{k=0}^{N-1} x_k e^{-j(2\pi/N)(n-m)k} = X(n-m)$$

(This is the frequency-shift property D4 of Table 5.1.) ●

The convolution property, D5(i), is arrived at in much the same way as for the Z-transform (Section 3.14), the difference being that instead of a series involving powers of z^{-1}, we now have sums in which powers of $e^{-j(2\pi/N)}$ appear. (It should be noted, however, that $X_1(n)X_2(n)$ is a term in a sequence. In the third column of Table 5.1 it is implied that $n = 0, 1, 2, \ldots$, whichever property is considered.) In the case of the DFT, as for the continuous transform, we have the symmetrical result, property D5(ii), where multiplication in the *time*-domain appears.

Properties D6 and D7, concerning a spectrum of real even and real odd sequences, are analogous to special forms of the Fourier trigonometric series and of the Fourier transform applicable in such cases. These properties are the subject of Problem 5.6.

Worked Example 5.6 (Complex Signals)

The most general form of a complex signal can be written $z(k) = x(k) + jy(k)$, (in which we make no assumption that either real $x(k)$ or real $y(k)$ exhibits any form of symmetry.) Find the DFT $X(n)$ in trigonometric form, making use of properties D6 and D7.

Solution

Any function can be written as a sum of even and odd functions. In particular,

$$x(k) = \tfrac{1}{2}[x(k) + x(-k)] + \tfrac{1}{2}[x(k) - x(-k)]$$

(Also, periodicity is implied: we could write

$$x(k) = \tfrac{1}{2}[x(k) + x(N - k)] + \tfrac{1}{2}[x(k) - x(N - k)]$$

instead of the previous equation.) Replacing k by $-k$ in the bracketed expressions, we see that the first is unaltered in value (and so is even) whereas the second would be the subject of a change in sign (and so is odd). Similarly,

$$y(k) = \tfrac{1}{2}[y(k) + y(-k)] + \tfrac{1}{2}[y(k) - y(-k)]$$
$$= \tfrac{1}{2}[y(k) + y(N - k)] + \tfrac{1}{2}[y(k) - y(N - k)]$$

We may then apply properties D6 and D7 to the four terms in question.

The result (as should be verified) is the equation

$$X(n) = \sum_{k=0}^{N-1} \left[\left\{ x_k \cos\left[\left(\frac{2\pi}{N}\right)nK \right] + y_k \sin\left[\left(\frac{2\pi}{N}\right)nK \right] \right\} \right.$$
$$\left. - j \left\{ x_k \sin\left[\left(\frac{2\pi}{N}\right)nK \right] - y_k \cos\left[\left(\frac{2\pi}{N}\right)nK \right] \right\} \right] \quad (5.16)$$

(This could, of course, have been obtained immediately from the definition of the DFT.) Equation (5.16) has implications to which we shall return in Section 5.9. ●

5.6 DISCRETE CORRELATION

The matter of identifying the parameters of a linear system by examining the transform of the system description was first raised in Worked Example 1.8 in the context of the Laplace transform. Subsequently (in Section 3.10) we examined how the transfer function of a system could be identified in terms of the Z-transform from observed sampled inputs and outputs. In both cases the use of convolution properties is implicit, since $Y = HX$.

In practice, a system is usually affected by *noise* (internal or external disturbance). This is equivalent to additional, extraneous, input and so contaminates the output observations. Accordingly, a transfer function is sensitive to this source of error. The examination of stochastic (or random) processes is beyond the scope of this text, but it is desirable to mention some other quantities used in the attempt to optimize transfer-function identification.

We begin by defining the (continuous-time) *correlation integral* of two functions to be

$$\int_{-\infty}^{\infty} x_1(t')x_2(t+t')dt' = x_1 \star x_2 \quad (5.17)$$

which has a Fourier transform $X_1(f)\overline{X_2(f)}$. If $x_1(t) = x_2(t) = x(t)$, we refer to the *autocorrelation integral*, in which case the transform can be written $|X(f)|^2$. Otherwise, we use the term *cross-correlation*.

The corresponding discrete-time entity is referred to as the *discrete correlation sum*, usually defined by the equation

$$\sum_{m=0}^{N-1} x_1(m)x_2(k+m) = x_1 \star x_2 \quad (5.18)$$

whose DFT is $\overline{X}_1(n)X_2(n), n = 0, 1, 2, \ldots, N-1$. (The transforms are called *cross-spectra*.)

Equations (5.17) and (5.18) have the same sort of structure as the corresponding expressions for continuous and discrete convolution, and can be handled in similar ways. One important difference is that correlation is not commutative, i.e. $x_1 \star x_2 \neq x_2 \star x_1$, whereas the convolution expressions are symmetric and also commutative, $x_1 * x_2 = x_2 * x_1$. Another difference is that periodicity of observed sequences is implied when using convolution properties, which is not so in the case of correlation. Cross-correlation is used also to detect time-lags in experimental data (such as are observed in chemical processes), particularly where there is a random element. Related quantities are defined immediately after we have obtained the discrete form of Parseval's theorem in the next section. Correlation concepts and properties are not confined to analysis based on the Fourier transforms. They are equally accessible if we are using, for example, the Z-transform and commercial spectral analysers are programmed accordingly.

5.7 PARSEVAL'S THEOREM

In Problem 3.3 signal energy was the subject of the analogue version of Parseval's theorem:

$$E = \int_{-\infty}^{\infty} |x(t)|^2 \, dt = \int_{-\infty}^{\infty} |X(f)|^2 \, df \tag{5.19}$$

Worked Example 5.7

Obtain the discrete equivalent of Parseval's theorem given a sequence of N sampled values of a signal and of its spectrum.

Solution

We may begin by writing

$$E = \sum_{k=0}^{N-1} |x_k|^2 = \sum_{k=0}^{N-1} x_k \cdot \bar{x}_k$$

From property D2 of Table 5.1, substituting for $\bar{x}_k$ gives

$$E = \sum_{k=0}^{N-1} x_k \left[\frac{1}{N} \sum_{n=0}^{N-1} \bar{X}_n e^{-j(2\pi/N)nk} \right]$$

$$= \frac{1}{N} \sum_{n=0}^{N-1} \bar{X}_n \left[\sum_{k=0}^{N-1} x_k e^{-j(2\pi/N)nk} \right]$$

on changing the order of summation. By the defining equation (5.5) it follows that

$$E = \frac{1}{N} \sum_{n=0}^{N-1} \bar{X}_n X_n$$

whence

$$E = \sum_{k=0}^{N-1} |x_k|^2 = \frac{1}{N} \sum_{n=0}^{N-1} |X_n|^2 \qquad (5.20)$$

●

Energy, as given by (5.20), can be regarded as the sum over all frequency components of the quantity

$$S_x(n) = \frac{1}{N} X_n \bar{X}_n \qquad n = 0, 1, \ldots, N-1 \qquad (5.21)$$

This is called the *energy spectral density function*. From (5.18), it is the DFT of

$$R_x(k) = \frac{1}{N} \sum_{m=0}^{N-1} x(m)x(k+m) \qquad (5.22)$$

known as the *autocorrelation function*. (Texts give various alternative expressions for $R_x(k)$, but they are equivalent to that given in (5.22).)

5.8 A NOTE ON SAMPLING IN THE FREQUENCY DOMAIN, AND A FURTHER COMMENT ON WINDOW FUNCTIONS

In this chapter we started with the immediate establishment of finite sums, on the assumption that discrete values of both $x(t)$ and $X(f)$ were available. In Section 3.6 we considered in some detail the effects of sampling and subsequently windowing a continuous time signal on its spectrum. At each step time-domain multiplications were identified with frequency-domain convolutions, and the final illustration was of a spectrum of periodic nature showing the features associated with aliasing and spectral leakage, $X_{sw}(f)$.

It is not proposed here to rehearse the inversion process, but it is recommended that the reader produce a set of comparable diagrams showing the effect of sampling and windowing a transform ($X(f)$ or – preferably – $X_{sw}(f)$) and note what time-domain convolutions occur during the process. (The last paragraph of Section 3.6

stated at what intervals Δf the function $X_{sw}(f)$ should be sampled, and what length of window should be adopted.) This exercise will assist in understanding what are the underlying implications of using discrete data sets, those not being forthcoming from examination of (5.5) and (5.13).

In Section 3.6 we adopted a rectangular window. In Section 3.7 we described some of its disadvantages and gave, as an example of an alternative, the trigonometric Hanning window in both continuous and discrete forms. Two further examples are given here.

A discrete version of the lambda function might be used:

$$w(k) = \begin{cases} \dfrac{2k+1}{N} & k = 0, 1, \ldots, N/2 \\[3mm] 2 - \left(\dfrac{2k+1}{N}\right) & k = N/2 + 1, \ldots, N - 1 \end{cases}$$

(Sketch this.) This *triangular* window is an improvement on the rectangular window, as measured in terms of side-lobe levels and energy content, but is not as good as the Hanning window. A modification of the Hanning window, (3.7), is the *Hamming* window, in which the coefficients are altered slightly. The equation

$$w(k) = 0.54 - 0.46 \cos\left(\frac{2k+1}{N}\right)$$

is the definition usually adopted for the Hamming window. Among other windows, each with its own characteristics, are Kaiser–Bessel, Butterworth, and Chebyshev, and for further information about these the reader should refer to the literature dealing with practical applications, particularly to the design of digital filters.

5.9 COMPUTATIONAL EFFORT AND THE DISCRETE FOURIER TRANSFORM

In the earlier part of this chapter the question of simplifying the coefficient matrix of the DFT equations was addressed. In Worked Example 5.2, for the case $N = 4$, we examined in more detail in what combinations input data values appear in the computation of the DFT sums. (In Problem 5.4 this is extended to the case $N = 8$.) These examples demonstrated inherent inefficiency if the DFT is calculated directly from (5.5). The computational effort can be measured fairly exactly if we look at the trigonometric form of the DFT of a complex signal, obtained in Worked Example 5.6.

Having written z_k in terms of its real and imaginary parts x_k and y_k, and similarly replaced exponential factors by trigonometric functions, we obtained the transform X_n as given by (5.16), which represents N equations. If $n = 0$, we have $X_0 = \sum_{k=0}^{N-1} (x_k + jy_k)$, but for $n = 1, 2, \ldots, N - 1$ we must consider how many real multiplications are required. Each of these $N - 1$ equations is the sum of N terms, and each term includes four products (of x_k or y_k and a trigonometric function). It follows that, in principle, there are $4N(N - 1)$ real products to be found in the computation of an N-point DFT, using direct methods. (In practice the number will be rather less, as some of the trigonometric terms will reduce to $0, \pm 1$, but the count $4N(N - 1)$ is a sufficiently good estimate.) We may show that the number of real additions is the same. In terms of machine time, we can ignore addition in comparison with multiplication (and, equally, ignore factors $\pm j$), so that

$$M_D = 4N(N - 1) \tag{5.23}$$

remains our measure of 'cost'.

We have indicated previously that with radix 2, $N = 2^p$, and p might be any integer in a range giving $10^3 < N < 10^5$, in a realistic situation. For example, $p = 12$ gives $N = 2^{12} = 4096$. Equation (5.23) then tells us that the number of real multiplications required is of the order $M_D = 67\,092\,480$.

It will be seen in Chapter 6 that if the potential for carrying out nested calculations is exploited, this can be reduced to $M_F = 98\,304$.

SUMMARY

In Chapter 5 we have applied numerical integration and series truncation to the Fourier exponential series pair to define the discrete Fourier transform and its inverse in the form of two finite sums, each representing a set of N equations, each equation being a linear combination of N sampled values. That conjugation of inputs and outputs allows the use of the transform sum to be used for inversion purposes also has been demonstrated.

Properties of the transform pair have been developed, including the duality property, and in the context of system indentification and energy analysis we have introduced the concept of correlation and produced Parseval's theorem in its discrete form, with some extensions. Some further window functions have been introduced.

The cost-effectiveness of using the DFT definition directly has been queried throughout, and quantified in the final section.

PROBLEMS

5.1 Given $\{x_k = 1\}, n = 0, 1, \ldots, N - 1$, where $N = 8$, show that the DFT is given by

$$X_n = 8 \qquad n = 0$$
$$X_n = 0 \qquad n = 1, 2, \ldots, 7$$

by using (5.5). Verify, using (5.6), that inversion of $\{X_n\}$ recovers $\{x_k\}$ as defined.

5.2 Given $\{x_k = e^{j2\pi Tk/N}\}$, take $N = 8$ and $k = 0, 1, \ldots, 7$ and obtain the DFT $\{X_n\}$ in the form

$$X_n = \frac{1 - e^{j2\pi(T-n)}}{1 - e^{j\pi(T-n)/4}}$$

assuming $T \neq n$, for $n = 0, 1, 2, \ldots, 7$. (*Hint*: Use the expression for the sum of a geometric progression.)

 (i) Deduce that if T is an integer, not equal to n, then $X_n = 0$. What happens if $T = n$? Show that if $T = 1, X_n$ is real and $\mathbf{X}_n = (0, 8, 0, 0, 0, 0, 0, 0)^{\mathrm{T}}$.

(ii) Use the identity

$$\frac{e^{8j\theta} - 1}{e^{j\theta} - 1} = \frac{(e^{4j\theta} - e^{-4j\theta})e^{4j\theta}}{(e^{j\theta/2} - e^{-j\theta/2})e^{j\theta/2}}$$

to show that we may write X_n in the form

$$X_n = \frac{\sin[\pi(T-n)]}{\sin[\pi(T-n)/8]} \cdot e^{j7\pi(T-n)/8}$$

and that, if $T = 1.1$, the amplitude spectrum is given approximately by

$$|\mathbf{X}_n| = (0.74, 7.87, 0.89, 0.46, 0.34, 0.31, 0.33, 0.42)$$

Sketch the amplitude spectra for cases (i) and (ii). (Note that we have phase differences in case (ii) which results in a distribution of spectral energy between frequency components.)

5.3 In this question, use either the pair of equations (5.5) and (5.6) or the pair of equations (5.5) and (5.13), as you wish. (In either case, use the expression for the sum of a geometric progression if it helps.)

(i) Given $N = 5$ and $\{x_k\} = \{1, 1, 1, 1, 1\}^T, k = 0, 1, 2, 3, 4$, show that $\{X_n\} = \{1, 0, 0, 0, 0\}^T, n = 0, 1, 2, 3, 4$.

(ii) Given $N = 8$ and $\{x_k\} = \{0, 1, 1, 1, 1, 1, 0, 0\}^T, k = 0, 1, \ldots, 7$, show that $\{X_n\}$ can be found by evaluating

$$X_n = 2\cos\left(\frac{\pi n}{2}\right) + 2\cos\left(\frac{\pi n}{4}\right) + 1 \qquad n = 0, 1, \ldots, 7$$

(If you prefer to leave your result in exponential form, verify that the same numerical values are generated.)

Recover $\{x_k\}$ in full by inversion in part (i) of this question. In part (ii), recover $x_2 = 1$ and $x_6 = 0$.

5.4 With $N = 8$ and $W_8 = W = e^{-j\pi/4}$, show that the eight equations giving the DFT can be simplified to read:

$$\mathbf{X}_n = \begin{bmatrix} 1 & 1 & 1 & 1 & 1 & 1 & 1 & 1 \\ 1 & W^1 & W^2 & W^3 & -1 & -W^1 & -W^2 & -W^3 \\ 1 & W^2 & -1 & -W^2 & 1 & W^2 & -1 & -W^2 \\ 1 & W^3 & -W^2 & W^1 & -1 & -W^3 & W^2 & -W^1 \\ 1 & -1 & 1 & -1 & 1 & -1 & 1 & -1 \\ 1 & -W^1 & W^2 & -W^3 & -1 & W^1 & -W^2 & W^3 \\ 1 & -W^2 & -1 & W^2 & 1 & -W^2 & -1 & W^2 \\ 1 & -W^3 & -W^2 & -W^1 & -1 & W^3 & W^2 & W^1 \end{bmatrix} \mathbf{x}_k$$

in which $n = 0, 1, \ldots, 7$ and $k = 0, 1, \ldots, 7$. Expand and reorder the terms in these equations, to show that the constituent terms on the right-hand side are multiples of $(x_0 \pm x_4), (x_2 \pm x_6)$, $(x_1 \pm x_5)$ and $(x_3 \pm x_7)$. (Note that $(x_0, x_4, x_2, x_6, x_1, x_5, x_3, x_7)^T$ is the bit-reversed ordering of $\mathbf{x}_k$ – see Problem 3.1.)

5.5 By inverting $X_n e^{-j(2\pi/N)mn}$, in which m is some fixed integer, show that this is the transform of the signal $x(k - m)$.

5.6 (i) If $x(k)$ is an even function, then $x(k) - x(-k) = 0$. From the definition of the DFT, replacing the exponential function by its trigonometric equivalent, show that in this case

$$X_n = \sum_{k=0}^{N-1} x_k \cos\left(\frac{2\pi}{N}\right)nk$$

(ii) Similarly, if $x(k)$ is an odd function, then $x(k) + x(-k) = 0$.

Show that the spectrum is given by

$$X_n = -j \sum_{k=0}^{N-1} x_k \sin\left(\frac{2\pi}{N}\right)nk$$

5.7 Find the four-point DFT of $\mathbf{x}_k = (1, j, 2, 0)^{\mathrm{T}}$ using (5.16). Confirm that (5.5) gives the same result, namely

$$\mathbf{X}_n = (3 + j, 0, 3 - j, -2)^{\mathrm{T}}$$

5.8 Use (5.17) to write down the correlation integral $x_2 \star x_1$, and by means of a substitution show that $x_2 \star x_1 \neq x_1 \star x_2$.

5.9 Verify that Parseval's theorem

$$\sum_{k=0}^{N-1} |x_k|^2 = \frac{1}{N} \sum_{n=0}^{N-1} |x_n|^2$$

is satisfied using the data given and obtained in the preceding Problems 5.1, 5.2(ii), 5.3 and 5.7. (Note that your result will be approximate in the second of these.)

5.10 Tabulate values of the spectrum $X_n = 1 + j^n$, for $n = 0, 1, 2, \ldots, 7$ and use the inversion property D2 of Table 5.1 (i.e. equation (5.13)), to show that X_n is the transform of the signal

$$x_k = \begin{cases} 1 & \text{if } k = 0, 6 \\ 0 & \text{if } k = 1, 2, 3, 4, 5, 7 \end{cases}$$

5.11 Sketch the triangle function $x(t) = \Lambda[(t-1)/1], 0 \leqslant t \leqslant 2$. Tabulate sampled values, x_k, for $k = 0, 1, 2, \ldots, 7$, starting with $k = 0$ at $t = 0$ and taking sampling interval $T = \frac{1}{4}$. (Add these point values to your sketch, and note that $x(2)$ is not included.) Show that the DFT of x_k is the real spectrum

$$\mathbf{X}_n = \left(4, -\left(1 + \frac{1}{\sqrt{2}}\right), 0, -\left(1 - \frac{1}{\sqrt{2}}\right), 0, \right.$$

$$\left. -\left(1 - \frac{1}{\sqrt{2}}\right), 0, -\left(1 + \frac{1}{\sqrt{2}}\right)\right)^{\mathrm{T}}$$

(for $n = 0, 1, \ldots, 7$) and sketch it. Confirm that $E = 11/4$ is given by either time- or frequency-domain data.

5.12 (i) Given that $x_1(k) = \cos(k\pi/4)$ and $x_2(k) = \cos(k\pi/2), k = 0, 1, 2, \ldots, 7$, show that the tabulated values of $x(k) =$

$x_1(k) + x_2(k)$ are

$$\mathbf{x}_k = \left(2, \frac{1}{\sqrt{2}}, -1, -\frac{1}{\sqrt{2}}, 0, -\frac{1}{\sqrt{2}}, -1, \frac{1}{\sqrt{2}} \right)^{\mathrm{T}}$$

and that this has a DFT

$$\mathbf{X}_n = (0, 4, 4, 0, 0, 0, 4, 4)^{\mathrm{T}} \qquad n = 0, 1, \ldots, 7$$

(ii) If a spectrum is the sequence

$$\mathbf{X}_n = (0, 4, 0, 0, 0, 0, 0, 4)^{\mathrm{T}} \qquad n = 0, 1, \ldots, 7,$$

show by inversion that the corresponding time-domain sequence is $x_1(k)$, as defined in part (i) of this question. (Note the implication is that $x_2(k)$ has been filtered out.)

6

Simplification and factorization of the discrete Fourier transform matrix

In Chapter 5 we considered the set of equations

$$X(n) = \sum_{k=0}^{N-1} x(k) e^{-j(2\pi/N)nk} \qquad n = 0, 1, \dots, N-1$$

defining the DFT of the sequence $\{x(k)\}$ without any modification other than simplifying the exponential factors as far as possible. In matrix form, we can write

$$\mathbf{X}_n = \mathbf{M} \mathbf{x}_k \tag{6.1}$$

where M is an $N \times N$ matrix. Since the elements of M are exponentials, there are no zero elements whatsoever. Unless some signal values are zero, direct computation means that we have to evaluate N sums, each of N terms, and in Section 5.9 this was a costly operation. However, we have also seen, in Worked Example 5.2 and Problem 5.4, that pairs of signal values appear in the same combinations repeatedly in these simultaneous equations, and it is our purpose in this chapter to consider how we can ensure that any arithmetic operation which involves combining two numbers (whether signal values or not) is performed only once.

It will be shown that if the calculation is carried out in a number of stages, it is possible to achieve that objective, with very considerable saving in computational effort. Calculations can be broken down in such a way that at any particular stage (often called a *recursion*), the successive algebraic operations being carried out are such as to produce linear combinations each of just two known numbers. These linear combinations are themselves then paired and form the known input numbers to the next stage. If after the first stage we have outputs each combining two signal values, it follows that after the second stage each output represents a combination of four signal values. The staging process is repeated until each output number is, by implication, a weighted combination of all N signal values, in accordance with (6.1), and so is an element of the required transform X_n. (For example, there are three stages in the computation of the spectrum if $N = 8$.)

Breaking down the calculation in this way is known as *decimation*, and is equivalent to restructuring the problem by replacing the coefficient matrix M in (6.1) by a product of matrices. Each factor matrix is the coefficient matrix of the equations associated with one stage in the process, and if we are setting up linear combinations of two quantities it follows that the factor matrices are sparse, each row having only two non-zero elements.

We shall illustrate these remarks for the particular case $N = 8$. Also, for present purposes, we shall confine our attention to the process known as *decimation in time* (DIT), although at the end of Section 6.5 the idea of *decimation in frequency* (DIF) is introduced. In Worked Example 5.2 (and elsewhere in Chapter 5) we saw that, in the calculation of X_n, the signal value combinations were pairs from the bit-reversed ordering of x_k. That suggests that the permutation matrix E be included as a factor of M, so that at the first stage input values are ordered appropriately.

In Sections 6.6 and 6.7 we shall describe how the equations described using a factor matrix can be represented diagrammatically. Finally, we will estimate how many real multiplications are needed if the transform problem is restructured along these lines.

6.1 THE COEFFICIENT MATRIX FOR AN EIGHT-POINT DISCRETE FOURIER TRANSFORM

With $N = 8$, we may put

$$W = W_8 = e^{-j(2\pi/8)} = e^{-j\pi/4} \tag{6.2}$$

and then the coefficient matrix in (6.1), unsimplified, is given by the equation

$$
M = \begin{bmatrix}
W^0 & W^0 & W^0 & W^0 & W^0 & W^0 & W^0 & W^0 \\
W^0 & W^1 & W^2 & W^3 & W^4 & W^5 & W^6 & W^7 \\
W^0 & W^2 & W^4 & W^6 & W^8 & W^{10} & W^{12} & W^{14} \\
W^0 & W^3 & W^6 & W^9 & W^{12} & W^{15} & W^{18} & W^{21} \\
W^0 & W^4 & W^8 & W^{12} & W^{16} & W^{20} & W^{24} & W^{28} \\
W^0 & W^5 & W^{10} & W^{15} & W^{20} & W^{25} & W^{30} & W^{35} \\
W^0 & W^6 & W^{12} & W^{18} & W^{24} & W^{30} & W^{36} & W^{42} \\
W^0 & W^7 & W^{14} & W^{21} & W^{28} & W^{35} & W^{42} & W^{49}
\end{bmatrix}
$$

In Problem 5.4 the reader was asked to simplify this, which is achieved by noting that, from (6.2), $W^0 = 1$, $W^4 = -1$, and $W^8 = 1$ so that all elements of M can be replaced by ± 1, $\pm W^1$, $\pm W^2$ and $\pm W^3$. We repeat the result given in that question:

$$
M = \begin{bmatrix}
1 & 1 & 1 & 1 & 1 & 1 & 1 & 1 \\
1 & W^1 & W^2 & W^3 & -1 & -W^1 & -W^2 & -W^3 \\
1 & W^2 & -1 & -W^2 & 1 & W^2 & -1 & -W^2 \\
1 & W^3 & -W^2 & W^1 & -1 & -W^3 & W^2 & -W^1 \\
1 & -1 & 1 & -1 & 1 & -1 & 1 & -1 \\
1 & -W^1 & W^2 & -W^3 & -1 & W^1 & -W^2 & W^3 \\
1 & -W^2 & -1 & W^2 & 1 & -W^2 & -1 & W^2 \\
1 & -W^3 & -W^2 & -W^1 & -1 & W^3 & W^2 & W^1
\end{bmatrix}
$$

$$(6.3)$$

It is possible to proceed (as in Problem 5.4) to obtain the spectrum from this and decide retrospectively how the calculation might have been done more effectively. Instead, we shall propose a factorization of M and show what happens as the factor matrices are applied in succession. As said, we will accommodate bit-reversal by including the permutation matrix E and consider the restructuring of (6.1) in the form

$$
\mathbf{X}_n = M\mathbf{x}_k = M_8 M_4 M_2 E\mathbf{x}_k \tag{6.4}
$$

in which all four factors are 8×8 matrices. The matrices M_2, M_4 and M_8 are associated with three stages of calculation, and it will be seen that the subscripts refer to the number of signal values x_k which are combined in the outputs at each stage.

6.2 THE PERMUTATION MATRIX AND BIT-REVERSAL

Bit-reversal was discussed in Section 3.1 and the permutation matrix for the case $N = 8$ was introduced in Problem 3.1(ii), so from (6.4) we can say immediately that the input to the first stage will be

$$
E\mathbf{x}_k =
\begin{bmatrix}
1 & 0 & 0 & 0 & 0 & 0 & 0 & 0 \\
0 & 0 & 0 & 0 & 1 & 0 & 0 & 0 \\
0 & 0 & 1 & 0 & 0 & 0 & 0 & 0 \\
0 & 0 & 0 & 0 & 0 & 0 & 1 & 0 \\
0 & 1 & 0 & 0 & 0 & 0 & 0 & 0 \\
0 & 0 & 0 & 0 & 0 & 1 & 0 & 0 \\
0 & 0 & 0 & 1 & 0 & 0 & 0 & 0 \\
0 & 0 & 0 & 0 & 0 & 0 & 0 & 1
\end{bmatrix}
\begin{bmatrix}
x_0 \\ x_1 \\ x_2 \\ x_3 \\ x_4 \\ x_5 \\ x_6 \\ x_7
\end{bmatrix}
=
\begin{bmatrix}
x_0 \\ x_4 \\ x_2 \\ x_6 \\ x_1 \\ x_5 \\ x_3 \\ x_7
\end{bmatrix}
$$

6.3 THE OUTPUT FROM FOUR TWO-POINT DISCRETE FOURIER TRANSFORMS

The output from the first stage should provide us with linear combinations of bit-reversed pairs such as $(x_0 + x_4)$ and $(x_0 - x_4)$. We define the matrix M_2 as follows:

$$
M_2 =
\begin{bmatrix}
A & & & & \\
 & & & \text{zeros} & \\
 & A & & & \\
 & & A & & \\
\text{zeros} & & & & \\
 & & & & A
\end{bmatrix}
\tag{6.5}
$$

in which A is the 2×2 matrix

$$
A = \begin{pmatrix} 1 & W^0 \\ 1 & W^4 \end{pmatrix} = \begin{pmatrix} 1 & 1 \\ 1 & -1 \end{pmatrix}
\tag{6.6}
$$

and all other elements of M_2 are zero. This will give the outputs from the first stage in the form

$M_2(E\mathbf{x}_k)$

$$= \begin{bmatrix} 1 & 1 & & & & & & \\ 1 & -1 & & & \text{zeros} & & & \\ & & 1 & 1 & & & & \\ & & 1 & -1 & & & & \\ & & & & 1 & 1 & & \\ & \text{zeros} & & & 1 & -1 & & \\ & & & & & & 1 & 1 \\ & & & & & & 1 & -1 \end{bmatrix} \begin{bmatrix} x_0 \\ x_4 \\ x_2 \\ x_6 \\ x_1 \\ x_5 \\ x_3 \\ x_7 \end{bmatrix} = \begin{bmatrix} x_0 + x_4 \\ x_0 - x_4 \\ x_2 + x_6 \\ x_2 - x_6 \\ x_1 + x_5 \\ x_1 - x_5 \\ x_3 + x_7 \\ x_3 - x_7 \end{bmatrix}$$

These eight outputs, paired, can be regarded as the outputs from four two-point DFTs, and form the inputs to the second stage during which *they* will be combined. If we continued to write all results in terms of x_k then the development of this process would be obscured. Accordingly we introduce new notation and define

$$\begin{bmatrix} f_0 \\ f_1 \\ g_0 \\ g_1 \\ \hat{f}_0 \\ \hat{f}_1 \\ \hat{g}_0 \\ \hat{g}_1 \end{bmatrix} = \begin{bmatrix} x_0 + x_4 \\ x_0 - x_4 \\ x_2 + x_6 \\ x_2 - x_6 \\ x_1 + x_5 \\ x_1 - x_5 \\ x_3 + x_7 \\ x_3 - x_7 \end{bmatrix} \tag{6.7}$$

The reason for defining M_2 as we did should be fairly clear. Subsequent stages involve complex multipliers and larger groupings of signal values, and those groupings might not be easily identifiable, even in the case $N = 8$. In other words, the required structure of the factor matrices M_4 and M_8 is not immediately discernible (see Problem 5.4 again). In this chapter we shall justify the forthcoming definitions of M_4 and M_8 only by showing that they lead to the correct results. However, in Chapter 7 a formal algebraic method of decimation will be described, from which the factor matrices emerge.

6.4 THE OUTPUT FROM TWO FOUR-POINT DISCRETE FOURIER TRANSFORMS

In the second stage, values of x_k appear in the output in combinations of four, some of which are multiplied by a coefficient W^2. We define the 8×8 matrix M_4 by the equation

$$M_4 = \begin{bmatrix} B & 0 \\ 0 & B \end{bmatrix} \tag{6.8}$$

in which B is the 4×4 matrix

$$B = \begin{bmatrix} 1 & 0 & W^0 & 0 \\ 0 & 1 & 0 & W^2 \\ 1 & 0 & W^4 & 0 \\ 0 & 1 & 0 & W^6 \end{bmatrix} \tag{6.9}$$

in which we can put $W^4 = -1$ and $W^6 = -W^2$. Outputs from the second stage are then given by forming the product

$$M_4(M_2 \mathbf{E} \mathbf{x}_k) = \begin{bmatrix} 1 & 0 & 1 & 0 & & & & \\ 0 & 1 & 0 & W^2 & & \mathbf{0} & & \\ 1 & 0 & -1 & 0 & & & & \\ 0 & 1 & 0 & -W^2 & & & & \\ & & & & & & & \\ & \mathbf{0} & & & \text{as above, left} & & \\ & & & & & & & \end{bmatrix} \begin{bmatrix} f_0 \\ f_1 \\ g_0 \\ g_1 \\ \hat{f}_0 \\ \hat{f}_1 \\ \hat{g}_0 \\ \hat{g}_1 \end{bmatrix}$$

We again introduce further notation to describe the results, defining $\mathbf{F}$ and $\mathbf{G}$ so that their elements are the output elements

$$\begin{bmatrix} F_0 \\ F_1 \\ F_2 \\ F_3 \\ G_0 \\ G_1 \\ G_2 \\ G_3 \end{bmatrix} = \begin{bmatrix} f_0 + g_0 \\ f_1 + W^2 g_1 \\ f_0 - g_0 \\ f_1 - W^2 g_1 \\ \hat{f}_0 + \hat{g}_0 \\ \hat{f}_1 + W^2 \hat{g}_1 \\ \hat{f}_0 - \hat{g}_0 \\ \hat{f}_1 - W^2 \hat{g}_1 \end{bmatrix} \tag{6.10}$$

Reference to (6.7) shows that each element of $\mathbf{F}$ is a combination of the even-subscripted values x_0, x_2, x_4 and x_6, and that similarly $\mathbf{G}$ is a function of the odd-subscripted samples. The second recursion can therefore be interpreted as having given us the output from two four-point DFTs.

Before proceeding to the third and final stage, it might be illuminating to compare what we have done so far, having specified $N = 8$, with the results which direct computation would give in the cases where the actual value of N is either 2 or 4.

Worked Example 6.1

With $N = 2$, find the spectrum of the sequence (y_0, y_1) in which $y_0 = x_1$ and $y_1 = x_5$.

Solution

From the equation

$$X(n) = \sum_{k=0}^{N-1} x(k) e^{-j(2\pi/N)nk} \qquad n = 0, 1, \dots, N-1$$

with the appropriate changes we obtain

$$Y(n) = \sum_{k=0}^{1} y(k) e^{-j\pi nk} \qquad n = 0, 1$$

Coefficients are powers of $W = W_2 = e^{-j\pi} = -1$ and so

$$Y_0 = y_0 + y_1 = x_1 + x_5$$

and

$$Y_1 = y_0 + y_1 e^{-j\pi} = x_1 - x_5 \qquad \bullet$$

This might be a trivial example, but we have obtained the quantities $\hat{f}_0$ and $\hat{f}_1$ which appeared in (6.7) after the first recursion of the eight-point DFT, and it is worth noticing that in (6.5) and (6.6) the coefficient $W^4 = W_8^4 = (e^{-j\pi/4})^4$ appears, i.e. $W_2 = (W_8)^4$.

Worked Example 6.2

With $N = 4$, find the DFT of the sequence $(y_0, y_1, y_2, y_3) = (x_1, x_3, x_5, x_7)$.

Solution

In Worked Example 5.2 we obtained a four-point DFT and so, quoting that result in terms of y, we have

$$Y_0 = (y_0 + y_2) + (y_1 + y_3)$$
$$Y_1 = (y_0 - y_2) - j(y_1 - y_3)$$
$$Y_2 = (y_0 + y_2) - (y_1 + y_3)$$
$$Y_3 = (y_0 - y_2) + j(y_1 - y_3)$$

(These expressions were obtained with $W = W_4 = e^{-j\pi/2} = -j$.) If we now designate $\mathbf{y}$ to be the stated odd-subscripted sequence $\mathbf{x}$, then

$$Y_0 = (x_1 + x_5) + (x_3 + x_7)$$
$$Y_1 = (x_1 - x_5) - j(x_3 - x_7)$$
$$Y_2 = (x_1 + x_5) - (x_3 + x_7)$$
$$Y_3 = (x_1 - x_5) + j(x_3 - x_7)$$

$$(6.11)$$

With the last results in mind, consider the elements of $\mathbf{G}$ as given in equation (6.10). From (6.7), these can be written

$$G_0 = (x_1 + x_5) + \quad\ (x_3 + x_7)$$
$$G_1 = (x_1 - x_5) + W_8^2(x_3 - x_7)$$
$$G_2 = (x_1 + x_5) - \quad\ (x_3 + x_7)$$
$$G_3 = (x_1 - x_5) - W_8^2(x_3 - x_7)$$

Since

$$W_8^2 = (e^{-j\pi/4})^2 = e^{-j\pi/2} = W_4 = -j$$

we find that $\mathbf{G} = \mathbf{Y}$, as given in (6.11)   ●

These two examples have been offered as confirmation of the assertion that the stage one operator matrix M_2 replaces an eight-point DFT by four two-point DFTs, and that, with M_4 as defined, after two stages we have computed two four-point DFTs. If, when staging an N-point DFT calculation, in which $N = 2^p$, the results are consistent with those in which N is equal to $2^{p-1}, 2^{p-2}, \ldots$, then there is an implication that, inductively, one could devise a factorization appropriate to cases in which N equals $2^{p+1}, 2^{p+2}, \ldots$. (We shall pick up that point after Worked Example 6.3.)

6.5 THE OUTPUT FROM AN EIGHT-POINT DISCRETE FOURIER TRANSFORM

In the third and final stage of calculating the $N = 2^3$ DFT we set up linear combinations of the elements of $\mathbf{F}$ and $\mathbf{G}$ which emerged as outputs from the second stage. The associated matrix M_8 is defined by the equation

$$M_8 = \begin{bmatrix} I_4 & \begin{matrix} W^0 & 0 & 0 & 0 \\ 0 & W^1 & 0 & 0 \\ 0 & 0 & W^2 & 0 \\ 0 & 0 & 0 & W^3 \end{matrix} \\ I_4 & \begin{matrix} W^4 & 0 & 0 & 0 \\ 0 & W^5 & 0 & 0 \\ 0 & 0 & W^6 & 0 \\ 0 & 0 & 0 & W^7 \end{matrix} \end{bmatrix} \quad (6.12)$$

in which I_4 is the 4×4 unit matrix. Noting that $W^4 = e^{(-jn\pi/4)4} = -1$, and putting

$$C = \begin{bmatrix} W^0 & 0 & 0 & 0 \\ 0 & W^1 & 0 & 0 \\ 0 & 0 & W^2 & 0 \\ 0 & 0 & 0 & W^3 \end{bmatrix} \quad (6.13)$$

we may write

$$M_8 = \begin{bmatrix} I_4 & C \\ I_4 & -C \end{bmatrix}$$

which is another sparse matrix, there being only two non-zero elements in each row. If the matrices we have defined are indeed factors of M, then from (6.4) the third stage output (say **Q**), given by

$$Q = M_8(M_4 M_2 E x_k) = M_8 \begin{pmatrix} F \\ G \end{pmatrix}$$

should be the spectrum X_n.

From (6.10),

$$Q = \begin{bmatrix} 1 & 0 & 0 & 0 & 1 & 0 & 0 & 0 \\ 0 & 1 & 0 & 0 & 0 & W^1 & 0 & 0 \\ 0 & 0 & 1 & 0 & 0 & 0 & W^2 & 0 \\ 0 & 0 & 0 & 1 & 0 & 0 & 0 & W^3 \\ 1 & 0 & 0 & 0 & -1 & 0 & 0 & 0 \\ 0 & 1 & 0 & 0 & 0 & -W^1 & 0 & 0 \\ 0 & 0 & 1 & 0 & 0 & 0 & -W^2 & 0 \\ 0 & 0 & 0 & 1 & 0 & 0 & 0 & -W^3 \end{bmatrix} \begin{bmatrix} F_0 \\ F_1 \\ F_2 \\ F_3 \\ G_0 \\ G_1 \\ G_2 \\ G_3 \end{bmatrix}$$

$$Q = \begin{bmatrix} F_0 + & G_0 \\ F_1 + W^1 G_1 \\ F_2 + W^2 G_2 \\ F_3 + W^3 G_3 \\ F_0 - & G_0 \\ F_1 - W^1 G_1 \\ F_2 - W^2 G_2 \\ F_3 - W^3 G_3 \end{bmatrix} \tag{6.14}$$

Worked Example 6.3

Verify that Q_3, given by (6.14), is X_3 in the spectrum $\mathbf{X}_n = \mathbf{M}\mathbf{x}_k$.

Solution

From (6.14)

$$Q_3 = F_3 + W^3 G_3,$$

and from (6.10) and (6.7)

$$Q_3 = (f_1 - W^2 g_1) + W^3(\hat{f}_1 - W^2 \hat{g}_1)$$
$$= ([x_0 - x_4] - W^2[x_2 - x_6])$$
$$\quad + W^3([x_1 - x_5] - W^2[x_3 - x_7])$$

Putting $W^5 = -W^1$, this gives

$$Q_3 = x_0 + W^3 x_1 - W^2 x_2 + W^1 x_3 - x_4 - W^3 x_5 + W^2 x_6 - W^1 x_7.$$

This is X_3 as given by $\mathbf{X}_n = \mathbf{M}\mathbf{x}_k$, using the matrix M given in (6.3). (Similarly, we can show $Q_n = X_n$ for all other values of n.) ●

The factorization process can be extended, as we said at the end of Section 6.4. For example, if $N = 2^4$, there would be a fourth stage to which the inputs are two eight-point DFTs. The form of the additional factor M_{16} can be deduced by examining the definitions of M_2, M_4 and M_8 in succession as given for $N = 8$ in this chapter, and considering what the four 16×16 factor matrices would look like. We shall not pursue that here, however, as in Chapter 7 we shall be developing methods in which the matrix factors actually appear as features of staged calculations.

It should be pointed out that although in this chapter (and in

Chapter 7) we are confining our attention to radix-2 methods, meaning that N is an integer power of 2, efficient computational methods can be generalized to cover situations where this is not the case.

Before leaving the matter of matrix factorization, we will say something about decimation in frequency. The matrix M in the DFT $X = Mx$ is symmetric, so that $M = M^T$. The transposition property applied to matrix products gives us

$$M = M^T = (M_8 M_4 M_2 E)^T = E^T M_2^T M_4^T M_8^T$$

The permutation matrix E is also symmetrical (again seen by inspection). Consequently, an alternative to (6.4), dropping subscripts n and k, is

$$X = E M_2^T M_4^T M_8^T x \qquad (6.15)$$

in which we have put $E^T = E$.

Whereas in the DIT process we input x in bit-reversed order to the first stage and obtained the DFT X in natural order after the third, the implication of (6.15) is that x is entered in natural order. Staging then involves eight-point, four-point and two-point DFTs in succession. The output after the third stage, $M_2^T M_4^T M_8^T x$, is the spectrum in bit-reversed order. Bit-reversal in the frequency domain is therefore required to obtain X in natural order. The algebraic basis for DIF also will be described in Chapter 7.

6.6 'BUTTERFLY' CALCULATIONS

In Chapter 7 we shall be demonstrating how to construct charts – known as signal flow graphs (SFGs) – which can be used to assist in the staged calculations of DFTs. Nested calculations are performed working from left to right across such a chart, recording the input and output values for each recursion at nodes on the diagram.

We have seen that applying a factor matrix to a vector of known values results in an output vector, each element of which is a linear combination of two of the incoming values. This is illustrated in Fig. 6.1. Suppose known values q_1 and q_2 are recorded at input nodes I_1 and I_2. Arrows show how they are to be weighted and combined at each output node. In other words, the value to be recorded at O_1 is $(a_1 q_1 + a_2 q_2)$ and at O_2 the output is $(b_1 q_1 + b_2 q_2)$. In an SFG for an N-point DFT there are N horizontal lines, at the left-hand end of which are recorded the N inputs (whether signal or spectrum values, bit-reversed or not).

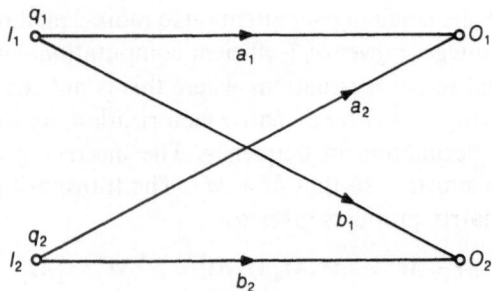

Fig. 6.1 A butterfly diagram.

We emphasize that in Fig. 6.1 there is *no* implication that the lines $I_1 O_1$ and $I_2 O_2$ are adjacent. For example, if we think of the lines as having 'addresses' $0, 1, \ldots, N-1$ from top to bottom, $I_1 O_1$ might be line '0' and $I_2 O_2$ line '3'.

6.7 'TWIDDLE' FACTORS

Returning to the results of matrix factoring, we see that the linear combinations to be produced at each stage are of particularly simple form and have the same structure. For example, from (6.7) we have output pairs such as

$$\hat{f}_0 = x_1 + x_5$$
$$\hat{f}_1 = x_1 - x_5$$

from (6.10), we have

$$F_1 = f_1 + W^2 g_1$$
$$F_3 = f_1 - W^2 g_1$$

and from (6.14), for example,

$$Q_3 = X_3 = F_3 + W^3 G_3$$
$$Q_7 = X_7 = F_3 - W^3 G_3 \tag{6.16}$$

The butterfly calculations all lead to outputs of the form $(q_1 \pm W^{(\text{power})} q_2)$. The diagrams all can be replaced by butterflies in which we are forming combinations $(q_1 \pm q_2)$ if any complex coefficient is accommodated beforehand. This is illustrated in Fig. 6.2, where we show the calculation of X_3 and X_7, as given in (6.16) using first a diagram like that of Fig. 6.1, and then a modification.

In Fig. 6.1(a), at the left F_3 and G_3 are at the output nodes from

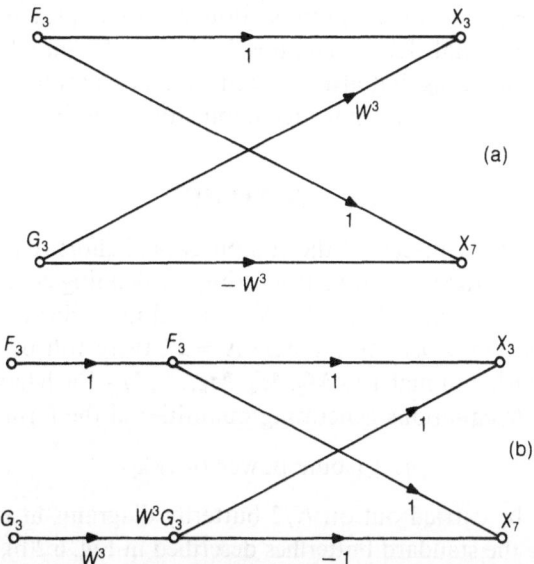

Fig. 6.2 Standardizing butterfly calculations.

some (two) previous butterflies and are then entered without modification at what are now the next input nodes. It is therefore necessary to multiply G_3 by $\pm W^3$ within the butterfly.

In Fig. 6.1(b) we have premultiplied the output G_3 so that the input to the butterfly is $W^3 G_3$, and it is now sufficient just to reverse its sign on the bottom line before computing X_7. If the SFG is stretched in this way, so that we deal with any complex numbers *between* butterflies, the complete chart has a much clearer and more symmetrical appearance. The complex multiplies are called *twiddle* factors. (Multiplication by $e^{j\theta}$ rotates a vector through angle θ.) For the eight-point DFT we are concerned only with W^1, W^2 and W^3 in this context, having already simplified all exponentials to values $\pm 1, \pm W^1, \pm W^2$ and $\pm W^3$. In an N-point DFT there are $(N/2 - 1)$ twiddle factors, (or $N/2$ if we regard $\pm W^0$ as such a factor).

Figures 6.1(a) and 6.1(b) are *fully labelled*. It is a fairly common procedure to indicate a factor $+1$ by an open arrow not actually labelled $+1$, and indeed the arrow might be omitted altogether as no change is made to the input node value along such a horizontal or diagonal. A similar convention is to interpret an unmarked solid arrow as sign-reversal – i.e. the node value is to be multiplied by -1. If this is clearly understood, the only arrows labelled in a graph will be those indicating multiplication by a complex number. In Chapter 7 we shall label diagrams in full in the initial stages, to assist understanding, but also show the 'cleaner' graph available for working if factors ± 1 are indicated implicitly rather than explicitly.

6.8 ECONOMIES

In Section 5.9 we discussed the computational effort required if the DFT is calculated without restructuring the defining equations, and obtained an estimator $M_D = 4N(N-1)$ real multiplications.

If decimation is used in the case $N = 2^p$, there will be $p = \log_2 N$ stages, and p factor matrices $M_2, M_4, M_8, \ldots, M_N$. Each factor matrix represents N equations generating quantities of the form

$$q_1 \pm (\text{some power of } W)q_2$$

which can be carried out on $N/2$ butterfly diagrams at each stage. If these are the standard butterflies described in Fig. 6.2(b), then each recursion implies that $N/2$ multiplications of two complex quantities are required, which in turn implies $4(N/2)$ *real* multiplications. The estimator of effort, should the factorization approach be adopted, can be taken as

$$M_F = 2N \log_2 N \tag{6.17}$$

real multiplications. Again this is an overestimate. In the first place, whatever the value of N, M_2 involves no complex multiplications. Secondly, in the matrices $M_4, M_8, \ldots$ a number of powers of W reduce to values ± 1.

Savings in effort are evident as soon as $N > 2$, and become increasingly, very substantially, significant.

SUMMARY

In Chapter 6 we have shown that factorization of the DFT matrix and nested calculation through decimation allows us to replace all equations by equations defining linear combinations of two quantities at any stage. No such combination is calculated more than once, resulting in a considerable saving in effort compared with that required by the direct use of the DFT equations.

In anticipation of the development of signal flow graphs in Chapter 7, we have introduced the principal graphical features of such charts (butterfly diagrams and twiddle factors).

PROBLEMS

In Problems 6.1–6.8, reference is made either to a worked example or to a problem from Chapter 5, in which (for $N = 4$ or for $N = 8$) the DFT of a time signal specified numerically was obtained directly. Reproduce those results by using the staged methods of calculation described in Sections 6.2–6.5 (i.e. using factor matrices).

6.1   Worked Example 5.1 ($N = 8$).
6.2   Worked Example 5.3 ($N = 4$).
6.3   Problem 5.3 ($N = 8$).
6.4   Problem 5.2(i) ($N = 8$).
6.5   Problem 5.3(ii) ($N = 8$).
6.6   Problem 5.7 ($N = 4$).
6.7   Problem 5.11 ($N = 8$).
6.8   Problem 5.12(i) ($N = 8$).
6.9   Confirm that $Q_2 = X_2$ and $Q_5 = X_5$, where $\mathbf{Q}$ is given by (6.14).
6.10 Show that M_2 is a symmetric matrix but that M_4 and M_8 are not.
6.11 With reference to (6.16), discussed in Section 6.7, sketch the butterfly diagrams for the remaining calculations associated with (6.14).
6.12 (i) On the same graph, sketch the functions $M_D = 4N(N - 1)$ and M_F as given in (6.17) for $2 < N < 64$.
      (ii) If you know how to handle limiting forms of indeterminate ratios, show that

$$\lim_{N \to \infty} \left(\frac{M_F}{M_D} \right) = 0$$

7

Fast Fourier transforms

It should be stated immediately that a fast Fourier transform (FFT) is not a 'new transform' but is an *algorithm* for the efficient calculation of the discrete Fourier transform. There are many such methods.

In Chapter 5 we introduced the DFT and considered its properties and inversion. In Chapter 6, largely expository, we investigated computational aspects, and the replacement of the DFT coefficient matrix by a product of sparse matrices was shown to result in a very considerable reduction in effort. In this concluding chapter, the object is to relate the ideas of Chapter 6 to a systematic form for DFT computation. The manner in which we achieve this objective will also show why the factor matrices $M_2, M_4, M_8, \ldots$ have the form they do. (Their definitions in (6.5), (6.8) and (6.12) might have appeared to be somewhat arbitrary, and were only justified by the fact that their application *does* generate the required spectrum.)

Although there are good routines for calculations which involve sparse matrices, the FFT does not incorporate those but directly parallels the structural similarity of factor matrices in an algorithm which can be illustrated graphically (and which also is easy to program.) We whall demonstrate this by decimating the DFT sums algebraically, to obtain the stage equations. That means that at each recursion, any sum is replaced by two sums in each of which the number of terms is halved. The process is repeated until all sums include just two signal values. The results which emerge from this are then considered in reverse order, starting with paired signal values, and recombined through staging until the final output is obtained in the form of a set of numbers satisfying the DFT equations – i.e. the elements of the spectrum. (This is equivalent to the explanation of staging given in the introduction to Chapter 6.) It will be seen that the whole process can be illustrated by constructing a signal flow graph (SFG) of which the essential features are butterfly diagrams and twiddle factors described in Sections 6.6 and 6.7.

We shall consider both decimation in time and decimation in frequency. (The most notable originators of these methods were Cooley, Tukey and Sande. Many alternative and modified FFTs have been developed over the last twenty or so years, with the names of their originators attached. What they have in common is the basic concept of decimation.) As said before, the discussion will be restricted to cases in which N is a power of 2. In Section 7.1 two possible ways of decimating (halving) a sum are outlined.

Once an algorithm has been translated into graphical form, it follows that the resulting SFG can also be used to invert a spectrum, because we showed in Section 5.4 that working with the complex conjugates of signals and spectra allows us to use the same algorithm for both transformation and inversion. This is discussed in detail in Section 7.6 in the context of DIT, and illustrated in Worked Example 7.5 using the decimation-in-frequency SFG.

7.1 FAST FOURIER TRANSFORM ALGORITHMS

In Chapter 6 an estimate was obtained of the reduced computational effort consequent on matrix factorization. The estimated number of real multiplications required using either of the FFTs developed in this chapter is the same, namely $M_F = 2N \log_2 N$.

Halving the DFT sum, referred to in the introduction, is the essential step in the recursive process and, in this context, is done in one of two ways. We have

$$X(n) = \sum_{k=0}^{N-1} x(k)W^{kn} \qquad n = 0, 1, \dots, N-1 \qquad (7.1)$$

in which $W = e^{-j(2\pi/N)}$.

If the decision is made to separate the terms in (7.1) in which k is even from those in which it is odd,

$$X(n) = \sum_{k=2r} x(k)W^{kn} + \sum_{k=2r+1} x(k)W^{kn} \qquad (7.2)$$

the eventual outcome is an algorithm known as *decimation in time*. Each of the sums in (7.2) is a sum of $N/2$ terms. All subsequent decimations are carried out on the same basis (for example, the sums are next split according to whether r is even or is odd). The DIT FFT will be described in detail in Sections 7.2–7.5.

An alternative division of the terms in (7.1) is to write

$$X(n) = \sum_{k=0}^{N/2-1} x(k)W^{kn} + \sum_{k=N/2}^{N-1} x(k)W^{kn} \qquad (7.3)$$

as a first step and in subsequent stages to distinguish between odd and even values of the *frequency*-domain subscript n. This leads to the *decimation in frequency* algorithm, which will be described in Section 7.7.

Whether considering DIT or DIF, we shall discuss the detail with $N = 8$. For higher values, we would put the appropriate value of N into either (7.2) or (7.3) and follow exactly the same procedures as will be described below.

7.2 DECIMATION IN TIME FOR AN EIGHT-POINT DISCRETE FOURIER TRANSFORM: FIRST STAGE

With $N = 8$, from (7.1) we have

$$X(n) = \sum_{k=0}^{7} x(k)W^{kn} \qquad n = 0, 1, \ldots, 7$$

in which

$$W = W_8 = e^{-j\pi/4}$$

Distinguishing between even and odd values of k, the equivalent of (7.2) written out in full is

$$X(n) = \{x(0)W^0 + x(2)W^{2n} + x(4)W^{4n} + x(6)W^{6n}\}$$
$$+ \{x(1)W^n + x(3)W^{3n} + x(5)W^{5n} + x(7)W^{7n}\} \qquad n = 0, 1, \ldots, 7$$

Note that W^n is a common factor in the second four-term sum. Let us define

$$F(n) = x(0)W^0 + x(2)W^{2n} + x(4)W^{4n} + x(6)W^{6n} = \sum_{r=0}^{3} x(2r)W^{2rn}$$

$$(7.4)$$

and

$$G(n) = x(1)W^0 + x(3)W^{2n} + x(5)W^{4n} + x(7)W^{6n} = \sum_{r=0}^{3} x(2r+1)W^{2rn}$$

$$(7.5)$$

It follows that

$$X(n) = F(n) + W^n G(n) \qquad n = 0, 1, \ldots, 7 \qquad (7.6)$$

Now although (7.6) represents eight equations, it is only necessary to calculate $F(n)$ and $G(n)$ for four values of n, because they are periodic functions.

Worked Example 7.1

Show that if $F(n)$ is as defined in (7.4), then $F(n + 4) = F(n)$.

Solution

We have

$$F(n + 4) = x(0)W^0 + x(2)W^{2(n+4)} + x(4)W^{4(n+4)} + x(6)W^{6(n+4)}$$

However, $W^8 = e^{-j2\pi} = 1$, and so $W^{16} = (W^8)^2 = 1$ and similarly $W^{24} = 1$, whence

$$F(n + 4) = x(0)W^0 + x(2)W^{2n} + x(4)W^{4n} + x(6)W^{6n} = F(n)$$

For example, $F(4) = F(0)$, $F(5) = F(1)$ and so on. ●

Both $F(n)$ and $G(n)$ have period 4. As a result it is sufficient to use (7.4) and (7.5) with $n = 0, 1, 2, 3$ only. This means that (7.6) represents the eight equations

$$\begin{aligned}
X(0) &= F(0) + G(0)W^0 \\
X(1) &= F(1) + G(1)W^1 \\
X(2) &= F(2) + G(2)W^2 \\
X(3) &= F(3) + G(3)W^3 \\
X(4) &= F(0) - G(0)W^0 \\
X(5) &= F(1) - G(1)W^1 \\
X(6) &= F(2) - G(2)W^2 \\
X(7) &= F(3) - G(3)W^3
\end{aligned} \qquad (7.7)$$

These expressions should be compared with the final output, (6.14), obtained when using the matrix-factorization approach, with particular reference to M_8 as given by (6.12).

7.3 THE SECOND STAGE: FURTHER PERIODIC ASPECTS

Equations (7.4) and (7.5) are the subject of the next recursion, and we distinguish between even and odd values of r by putting either

$r = 2p$ or $r = 2p + 1$. From (7.4), explicitly,

$$F(n) = \{x(0)W^0 + x(4)W^{4n}\} + \{x(2)W^{2n} + x(6)W^{6n}\}$$

Here, W^{2n} is a common factor in the second sum. We now define (for $k = 2r = 2(2p) = 4p$),

$$f(n) = x(0)W^0 + x(4)W^{4n} = \sum_{p=0}^{1} x(4p)W^{4pn} \qquad (7.8)$$

and (for $k = 2r = 2(2p + 1) = 4p + 2$)

$$g(n) = x(2)W^0 + x(6)W^{4n} = \sum_{p=0}^{1} x(4p+2)W^{4pn} \qquad (7.9)$$

and then

$$F(n) = f(n) + W^{2n}g(n) \qquad n = 0, 1, 2, 3 \qquad (7.10)$$

Reordering (7.5)

$$G(n) = \{x(1)W^0 + x(5)W^{4n}\} + \{x(3)W^{2n} + x(7)W^{6n}\}$$

If we define (for $k = 2r + 1 = 2(2p) + 1 = 4p + 1$)

$$\hat{f}(n) = x(1)W^0 + x(5)W^{4n} = \sum_{p=0}^{1} x(4p+1)W^{4pn} \qquad (7.11)$$

and (for $k = 2r + 1 = 2(2p + 1) + 1 = 4p + 3$)

$$\hat{g}(n) = x(3)W^0 + x(7)W^{4n} = \sum_{p=0}^{1} x(4p+3)W^{4pn} \qquad (7.12)$$

it follows that

$$G(n) = \hat{f}(n) + W^{2n}\hat{g}(n) \qquad n = 0, 1, 2, 3 \qquad (7.13)$$

The amount of computation is again halved because the newly introduced function $f(n)\cdots\hat{g}(n)$ are periodic, and the period is 2. Note the similarity in structure of (7.6)–(7.13).

Worked Example 7.2

Show that $\hat{g}(n + 2) = \hat{g}(2)$.

Solution

From (7.12), since $W^8 = 1$,

$$\hat{g}(n + 2) = x(3) + x(7)W^{4(n+2)} = \hat{g}(n). \qquad \bullet$$

Equations (7.8), (7.9), (7.11) and (7.12), with $n = 0, 1$ only, therefore provide all the information needed to find $F(n)$ and $G(n)$ from (7.10) and (7.13). Putting $f(2) = f(0), f(3) = f(1)$, etc., and noting that $W^4 = -1$, (7.10) gives

$$
\begin{aligned}
F(0) &= f(0) + g(0) \\
F(1) &= f(1) + g(1)W^2 \\
F(2) &= f(0) - g(0) \\
F(3) &= f(1) - g(1)W^2
\end{aligned}
\tag{7.14}
$$

Similarly, from (7.13),

$$
\begin{aligned}
G(0) &= \hat{f}(0) + \hat{g}(0) \\
G(1) &= \hat{f}(1) + \hat{g}(1)W^2 \\
G(2) &= \hat{f}(0) - \hat{g}(0) \\
G(3) &= \hat{f}(1) - \hat{g}(1)W^2
\end{aligned}
\tag{7.15}
$$

The coefficients in these equations should be compared with the elements of factor matrix M_4, defined in (6.8), and the output from two four-point DFTs, (6.10), is consistent with equations (7.14) and (7.15) if the quantities $f(n)$ and $\hat{g}(n)$ are the same.

7.4 THE THIRD STAGE

In general we would repeat the process by next distinguishing between $p = 2q$ and $p = 2q + 1$, and halve the sums defining $f(n)$ and $\hat{g}(n)$. It is not necessary in this case as we have now arrived at pairings of signal values, and so the third and last recursion results in one-term 'sums', $x_0, x_1, \ldots, x_7$. Explicitly, we have, from (7.8), (7.9), (7.11) and (7.12)

$$
\begin{aligned}
f(0) &= x_0 + x_4 \\
f(1) &= x_0 - x_4 \\
g(0) &= x_2 + x_6 \\
g(1) &= x_2 - x_6 \\
\hat{f}(0) &= x_1 + x_5 \\
\hat{f}(1) &= x_1 - x_5 \\
\hat{g}(0) &= x_3 + x_7 \\
\hat{g}(1) &= x_3 - x_7
\end{aligned}
\tag{7.16}
$$

The breakdown is therefore fully consistent with the definition of M_2 given in (6.6) and the output from the application of M_2 of the bit-reversed signal vector, (6.7).

7.5 CONSTRUCTION OF A FLOW GRAPH

Equations (7.16) give the outputs from four two-point DFTs and there are no complex coefficients. We take these as the starting point to obtain the SFG for an eight-point DIT FFT. We shall use the standard butterfly diagram (Fig. 6.2(b)) throughout.

As the signal values are combined in bit-reversed pairs, we enter them in that order on the left of an eight-line diagram, of which the top four lines are associated with even-subscripted x_k, as shown in Fig. 7.1.

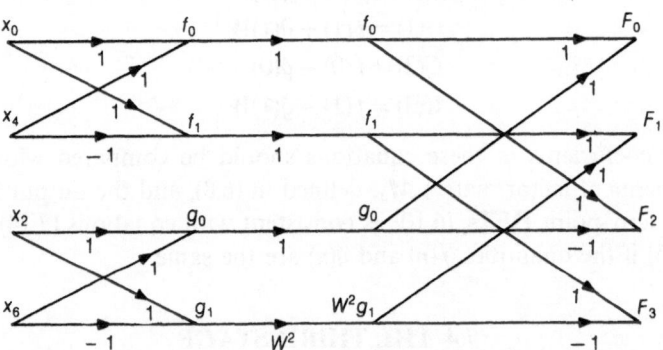

Fig. 7.1 The start of the FFT, even-subscripted inputs.

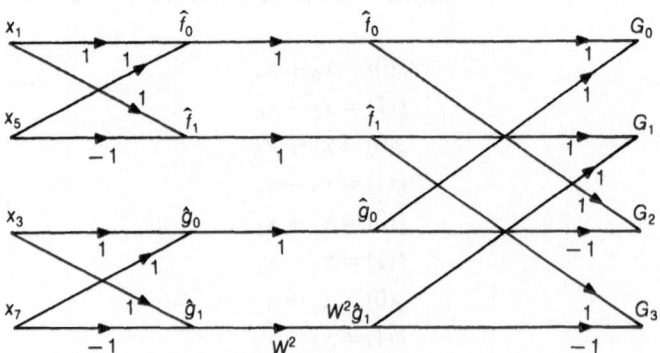

Fig. 7.2 Initial treatment of odd-subscripted inputs.

The first two butterflies give the first four quantities in (7.16) at the output nodes. Now refer to (7.14). This represents a four-point DFT in which $g(1)$ has a coefficient $W^2 = -j$. On the bottom line of Fig. 7.1 we have indicated a twiddle factor W^2, to give the relevant value. The next two butterflies again form the sums and differences of two inputs. (Notice that the butterflies now combine values from nodes which are *not* on adjacent lines, and confirm that the outputs F_n are as given by (7.14).)

To compute $\hat{f}(0)$, $\hat{f}(1)$, $\hat{g}(0)$ and $\hat{g}(1)$ from (7.16) we enter the odd-subscripted values of x_k, again in bit-reversed order, and the lower four lines of our eight-point DIT FFT chart will be as shown in Fig. 7.2. Having recorded those quantities at the output nodes of the first butterflies, the first stage is completed as far as the *lower* half of the graph is concerned. We proceed to the four-point DFT defined by (7.15), noting that in the computation of $\mathbf{G}$, $\hat{g}(1)$ is to be multiplied by W^2 (hence the premultiplier on the bottom line).

On the complete (eight-line) diagram there are now recorded the eight numbers $\mathbf{F}$ and $\mathbf{G}$ as outputs from the two four-point DFTs, and it remains to compute the eight-point DFT giving the required spectrum $X(n)$. From (7.7) we see that G_1, G_2 and G_3 on our chart should be multiplied by W^1, W^2 and W^3, respectively, to obtain the numbers required at the relevant input nodes of the final butterflies. In Fig. 7.3, therefore, twiddle factors appear on the last three lines.

We see that the final output, $X(n)$, appears in natural order, $n = 0, 1, 2, \ldots, 7$. In the complete SFG we have three stages, and there are four butterfly calculations at each stage. In the SFG for an N-point FFT, if $N = 2^p$ there will be p stages, each of $N/2$ butterflies. Figure 7.4 is a 'blank' chart showing just the multipliers required, further simplified by using an unmarked black arrow to indicate a factor (-1), and omitting altogether the arrows indicating $(+1)$. This illustrates the simplification of the SFG in respect of factors ± 1 which was described at the end of Section 6.8. The reader might prefer to retain all arrows and explicit labels when first making use of a SFG, but after some experience it is sufficient to remember that, of the two outputs from any standard butterfly, the sum of the two inputs is recorded at the upper node and their difference at the lower node.

Depending on the nature of the inputs, twiddle factors can be left in exponential form, or it might be more convenient to use the trigonometric or Cartesian equivalent. This also depends on whether the calculations are being done manually or using a programmed

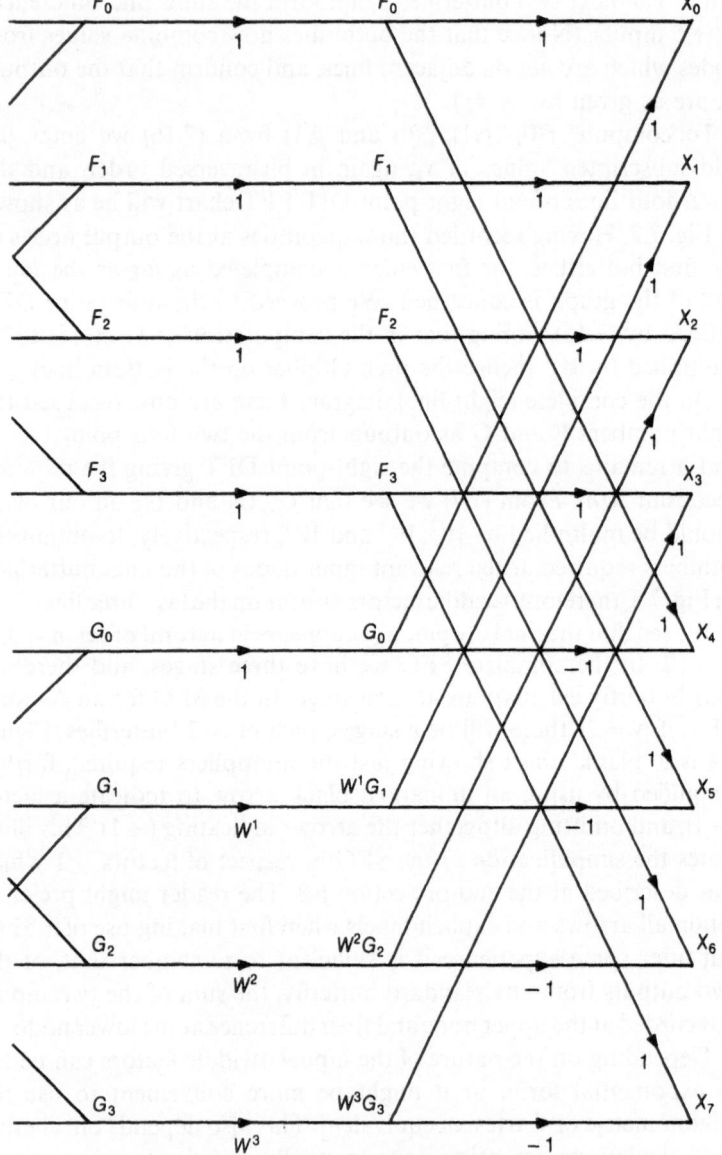

Fig. 7.3 The final stage of the DIT eight-point DFT.

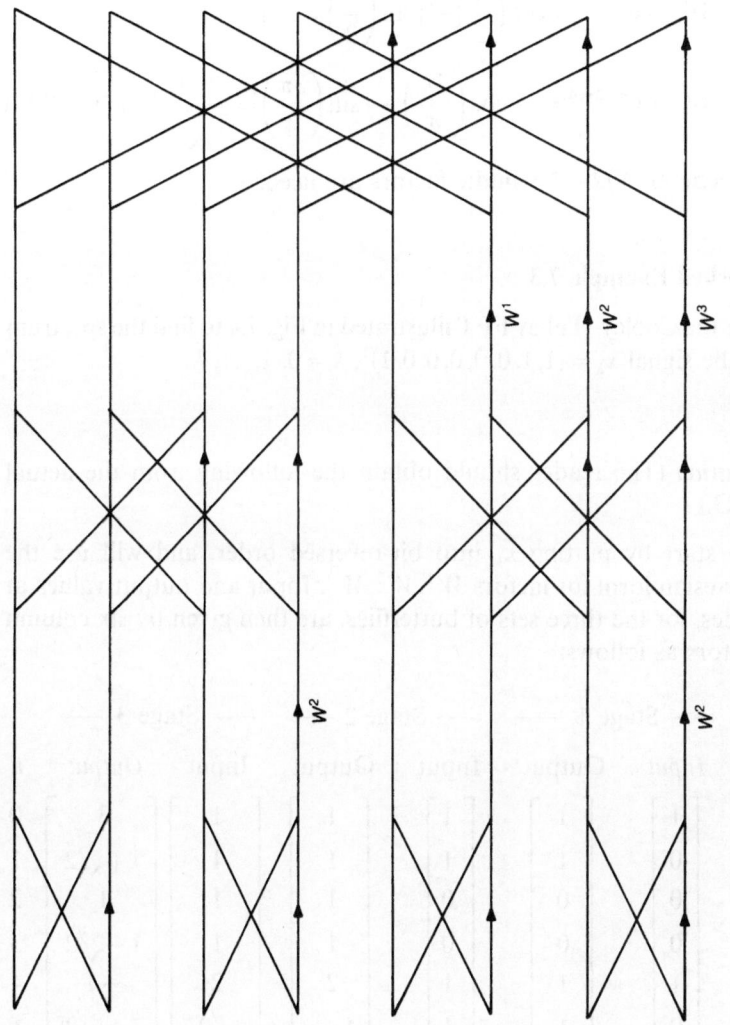

Fig. 7.4 The SFG for an eight-point DIT FFT.

version of the SFG. With $N = 8$ we have

$$W^1 = e^{-j\pi/4} = \cos\left(\frac{\pi}{4}\right) - j\sin\left(\frac{\pi}{4}\right) = \frac{1}{\sqrt{2}}(1 - j)$$

$$W^2 = e^{-j\pi/2} = \cos\left(\frac{\pi}{2}\right) - j\sin\left(\frac{\pi}{2}\right) = -j$$

$$W^3 = e^{-j3\pi/4} = -\cos\left(\frac{3\pi}{4}\right) - j\sin\left(\frac{3\pi}{4}\right) = -\frac{1}{\sqrt{2}}(1 + j) \quad (7.17)$$

In general, $N/2 - 1$ twiddle factors are needed.

Worked Example 7.3

Use the Cooley–Tukey FFT illustrated in Fig. 7.4 to find the spectrum of the signal $\mathbf{x}_k = (1, 1, 0, 0, 0, 0, 0, 1)^T$, $k = 0, 1, \ldots, 7$.

Solution (The reader should obtain the following from the actual SFG.)

We start by putting $\mathbf{x}_k$ into bit-reversed order, and will use the Cartesian form for factors W^1, W^2, W^3. Input and output values at nodes, for the three sets of butterflies, are then given by six column vectors as follows:

	← Stage 1 →		← Stage 2 →		← Stage 3 →		
k	*Input*	Output	Input	Output	Input	*Output*	n
0	1	1	1	1	1	3	0
4	0	1	1	1	1	$1+\sqrt{2}$	1
2	0	0	0	1	1	1	2
6	0	0	0	1	1	$1-\sqrt{2}$	3
1	1	1	1	2	2	-1	4
5	0	1	1	$1+j$	$\sqrt{2}$	$1-\sqrt{2}$	5
3	0	1	1	0	0	1	6
7	1	-1	j	$1-j$	$-\sqrt{2}$	$1+\sqrt{2}$	7

The stage 1 input is the bit-reversed $\mathbf{x}_k$ and the stage 3 output is $\mathbf{X}_n$

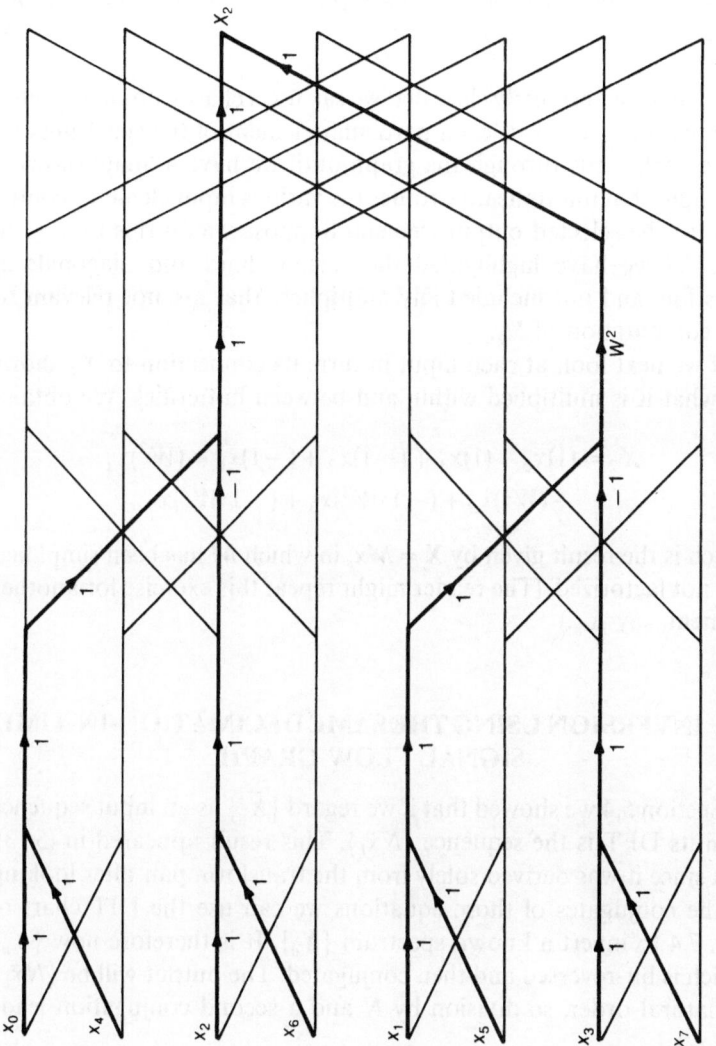

Fig. 7.5 The composition of X_2.

in natural order. (We may observe that $\sum|x_k|^2 = 3$ and $\sum|X_n|^2 = 24$, and Parseval's theorem is satisfied.) In all these vectors, the elements are ordered in the same way as in Figs 7.1, 7.2 and 7.3 from top to bottom as they appear on the SFG. They were obtained by working directly on the graph, and we have presented them in this form for clarity. ●

To make it absolutely clear that we can use a chart without reference to equations we can select a particular element of the spectrum and trace backwards through the graph until we have a 'map' showing through what multiplicative routes the various input elements contribute to the selected output element. Suppose we do this for X_2. In Fig. 7.5 we have highlighted the various lines and diagonals in question, and not included any multipliers that are not relevant to the computation of $\mathbf{X}_2$.

If we next look at each input in turn, its connection to X_2 shows by what it is multiplied within and between butterflies. We obtain

$$X_2 = (1)x_0 + (1)x_4 + (-1)x_2 + (-1)x_6 + (W^2)x_1$$
$$+ (W^2)x_5 + (-1 \cdot W^2)x_3 + (-1 \cdot W^2)x_7$$

which is the result given by $\mathbf{X} = \mathbf{M}\mathbf{x}$, in which M has been simplified but not factorized. (The reader might repeat this exercise for another element, say X_3.)

7.6 INVERSION USING THE SAME DECIMATION-IN-TIME SIGNAL FLOW GRAPH

In Section 5.4 we showed that if we regard $\{\bar{X}_n\}$ as an input sequence then its DFT is the sequence $\{N\bar{x}_k\}$. This result appeared in (5.13), and since it was derived solely from the transform pair after looking at the conjugates of those equations we can use the DIT chart of Fig. 7.4 to invert a known spectrum $\{X_n\}$. It is therefore now $\{X_n\}$ which is bit-reversed and then conjugated. The output will be $\{N\bar{x}_k\}$ in natural order, so division by N and a second conjugation leads to $\{x_k\}$.

Worked Example 7.4

Given that $N = 8$ and that a spectrum is

$$
\begin{bmatrix} X_0 \\ X_1 \\ X_2 \\ X_3 \\ X_4 \\ X_5 \\ X_6 \\ X_7 \end{bmatrix}
=
\begin{bmatrix} 3 \\ (1+\sqrt{2})j \\ 1 \\ (1-\sqrt{2})j \\ -1 \\ (1-\sqrt{2})j \\ 1 \\ (1+\sqrt{2})j \end{bmatrix}
$$

find the inverse.

Solution

We can present the values to be recorded at the SFG nodes in the form of vectors, as in Worked Example 7.3. Again we recommend verifying these results by working on a copy of the graph.

	← Stage 1 →		← Stage 2 →		← Stage 3 →		
n	Input	Output	Input	Output	Input	Output	k
0	3	2	2	4	4	$4(1-j)$	0
4	-1	4	4	4	4	$4(1-j)$	1
2	1	2	2	0	0	0	2
6	1	0	0	4	4	$4(1+j)$	3
1	$-(1+\sqrt{2})j$	$-2j$	$-2j$	$-4j$	$-4j$	$4(1+j)$	4
5	$-(1-\sqrt{2})j$	$-2\sqrt{2}j$	$-2\sqrt{2}j$	$2\sqrt{2}(1-j)$	$-4j$	$4(1+j)$	5
3	$-(1-\sqrt{2})j$	$-2j$	$-2j$	0	0	0	6
7	$-(1+\sqrt{2})j$	$2\sqrt{2}j$	$2\sqrt{2}$	$-2\sqrt{2}(1+j)$	$4j$	$4(1-j)$	7

In this, the stage 1 input is $\{X_n\}$ after bit-reversal and conjugation. The stage 3 output is $\{8\bar{x}_k\}$, and so we conclude that $x_0 = x_1 = x_7 = \frac{1}{2}(1 + j)$, $x_2 = x_6 = 0$, and $x_3 = x_4 = x_5 = \frac{1}{2}(1 - j)$. Similarly, any other FFT graph can be used for both transformation and inversion. ●

7.7 DECIMATION IN FREQUENCY FOR AN EIGHT-POINT DISCRETE FOURIER TRANSFORM

In Section 7.1 we suggested that another way of replacing an N-term summation by two sums of $N/2$ terms would be to halve their natural ordering. With $N = 8$, the equivalent of (7.3) is

$$X(n) = \sum_{k=0}^{3} x(k) W^{kn} + \sum_{k=4}^{7} x(k) W^{kn} \qquad n = 0, 1, \ldots, 7$$

where $W = W_8 = e^{-j\pi/4}$.

It is more convenient if both sums have the same limits. In the second, putting $k = m + 4$ gives

$$\sum_{k=4}^{7} x(k) W^{kn} = \sum_{m=0}^{3} x(m + 4) W^{(m+4)n}$$

The factor W^{4n} can be written before the summation sign, and we note that $W^{4n} = (e^{-j\pi})^n = (-1)^n$. Changing the dummy summation integer again, putting $m = k$, the eight-point DFT $X(n)$ can now be expressed

$$X(n) = \sum_{k=0}^{3} x(k) W^{kn} + (-1)^n \sum_{k=0}^{3} x(k + 4) W^{kn} \qquad n = 0, 1, \ldots, 7$$

$$(7.18)$$

Decimation in the *frequency* domain means that we are dividing the *output*, distinguishing between even and odd values of n (not k). For the cases $n = 2r$ and $n = 2r + 1$, from (7.18) we have

$$X(2r) = \sum_{k=0}^{3} [x(k) + x(k + 4)] W^{2kr} \qquad r = 0, 1, 2, 3 \qquad (7.19)$$

and

$$X(2r + 1) = \sum_{k=0}^{3} [x(k) - x(k + 4)] W^{k} \cdot W^{2kr} \qquad r = 0, 1, 2, 3 \qquad (7.20)$$

These can be thought of as two four-point DFTs if regarded as the transforms of sequences

$$\{y_k\} = \{x_k + x_{k+4}\}$$

and

$$\{y_k\} = \{(x_k - x_{k+4}) W^k\}$$

respectively, because $W_8^{2r} = W_4^r$. Subsequent decimations also represent DFTs.

We shall continue the process, but in terms of $\{x_k\}$ throughout.

The sums in (7.19) and (7.20) are halved in the next decimation and, where $\sum_{k=2}^{3}$ is seen, the substitution $k = m + 2$ will enable us to express all results as sums over $k = 0, 1$.

As an example, we will consider (7.19) with $r = 2p$, and write

$$X(4p) = \sum_{k=0}^{1} [x(k) + x(k+4)]W^{4kp} + \sum_{k=2}^{3} [x(k) + x(k+4)]W^{4kp}$$

$$(7.21)$$

Putting $k = m + 2$ in the second sum gives

$$\sum_{m=0}^{1} [x(m+2) + x(m+6)]W^{4(m+2)p}$$

in which $W^{8p} = 1$ has appeared. Replacing m by k we have, from (7.21),

$$X(4p) = \sum_{k=0}^{1} \{[x(k) + x(k+4)] + [x(k+2) + x(k+6)]\}W^{4kp}$$

In full (i.e. putting $k = 0$ and 1),

$$X(4p) = (x_0 + x_4) + (x_2 + x_6) + (x_1 + x_5)W^{4p}$$
$$+ (x_3 + x_7)W^{4p} \qquad p = 0, 1 \qquad (7.22)$$

Putting $r = 2p$ in (7.20) leads to

$$X(4p+1) = (x_0 - x_4) + (x_2 - x_6)W^2 + (x_1 - x_5)W^{4p+1}$$
$$+ (x_3 - x_7)W^{4p+3} \qquad p = 0, 1 \qquad (7.23)$$

and substituting $r = 2p + 1$ in (7.19) and (7.20) gives

$$X(4p+2) = (x_0 + x_4) - (x_2 + x_6) + (x_1 + x_5)W^{4p+2}$$
$$- (x_3 + x_7)W^{4p+2} \qquad p = 0, 1 \qquad (7.24)$$

and

$$X(4p+3) = (x_0 - x_4) - (x_2 - x_6)W^2 + (x_1 - x_5)W^{4p+3}$$
$$- (x_3 - x_7)W^{4p+5} \qquad p = 0, 1 \qquad (7.25)$$

The confirmation of these results is asked for in Problem 7.6.

The third decimation is to distinguish between cases $p = 0$ (even) and $p = 1$ (odd), but the consequences are sufficiently apparent from the preceding four equations and need not be made explicit. On first inspection, it might appear that four butterfly calculations should be carried out using bit-reversed inputs. This is a misleading impression as one could not then accommodate the required coefficients in subsequent stages without in effect reproducing the decimation in *time* SFG, which is not our intention. In several texts, the regrouping of terms in (7.22)–(7.25) is completed and the SFG constructed accordingly, with $\{x_k\}$ entered in natural order. This we leave for the reader to do, in Problem 7.12. Instead, we will illustrate how the SFG can be obtained from the factorized matrix, appropriately transposed.

The factor matrices defined in Chapter 6 were formally verified as they emerged naturally in the development of the DIT FFT (Sections 7.2, 7.3 and 7.4), and in Section 6.5 it was established that the product $M_2^T M_4^T M_8^T x_k$ will give X_n in bit-reversed order (if the permutation matrix E is omitted).

Transposing M_8, given by (6.12), the first stage output is

$$
M_8^T x_k = \begin{bmatrix}
1 & 0 & 0 & 0 & 1 & 0 & 0 & 0 \\
0 & 1 & 0 & 0 & 0 & 1 & 0 & 0 \\
0 & 0 & 1 & 0 & 0 & 0 & 1 & 0 \\
0 & 0 & 0 & 1 & 0 & 0 & 0 & 1 \\
1 & 0 & 0 & 0 & -1 & 0 & 0 & 0 \\
0 & W^1 & 0 & 0 & 0 & -W^1 & 0 & 0 \\
0 & 0 & W^2 & 0 & 0 & 0 & -W^2 & 0 \\
0 & 0 & 0 & W^3 & 0 & 0 & 0 & -W^3
\end{bmatrix}
\begin{bmatrix}
x_0 \\ x_1 \\ x_2 \\ x_3 \\ x_4 \\ x_5 \\ x_6 \\ x_7
\end{bmatrix}
$$

$$
= \begin{bmatrix}
(x_0 + x_4) \\
(x_1 + x_5) \\
(x_2 + x_6) \\
(x_3 + x_7) \\
(x_0 - x_4) \\
(x_1 - x_5)W^1 \\
(x_2 - x_6)W^2 \\
(x_3 - x_7)W^3
\end{bmatrix} \tag{7.26}
$$

With reference to Fig. 7.6, we see that if we enter $\{x_k\}$ in natural order on a graph and carry out an eight-point DFT at the *first* stage, the butterfly outputs, suitably postmultiplied by twiddle factors, provide the values obtained in (7.26).

These form the input node values at the next stage. Standard butterfly calculations applied to the stage 2 four-point DFTs, with postmultiplication by W^2 on the lowest line of each, gives the values $M_4^T(M_8^T x)$. These form the inputs to four two-point butterflies, and the final output is X_n in bit-reversed order. (Verification of these statements is asked for in Problem 7.7.)

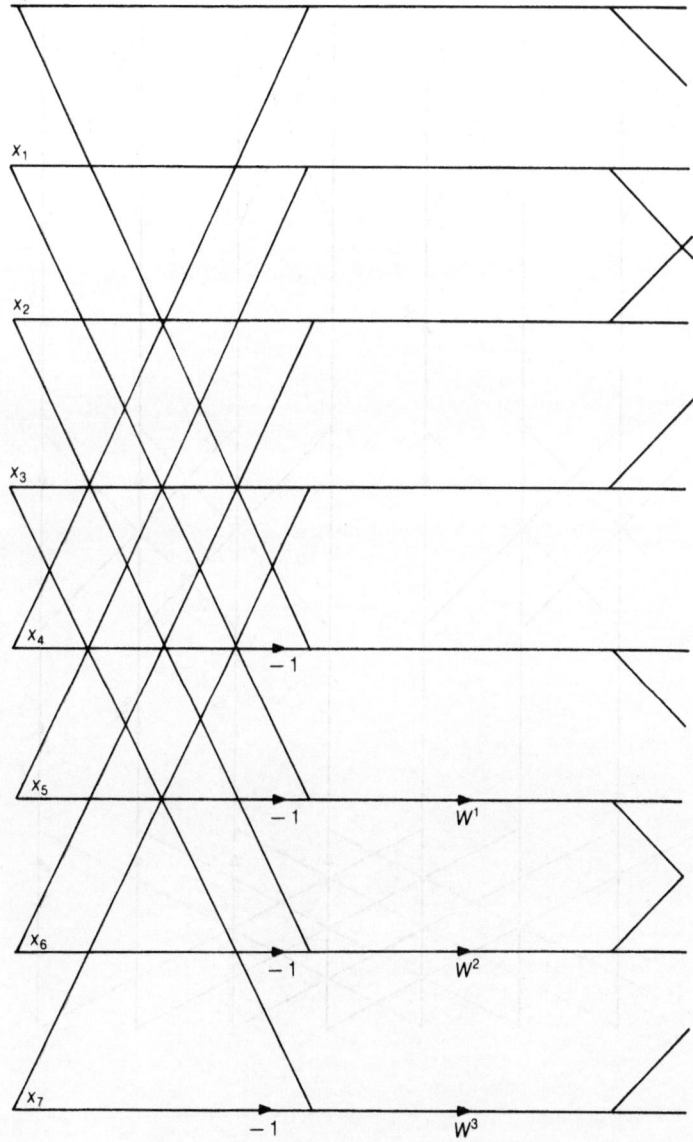

Fig. 7.6 The first stage of a DIF chart.

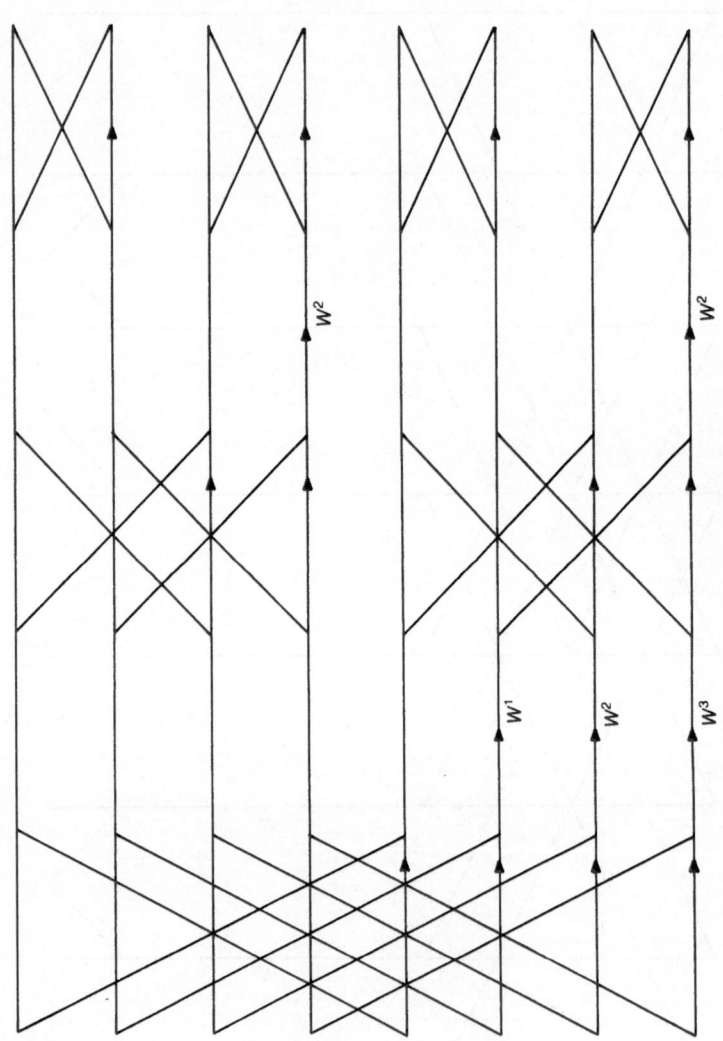

Fig. 7.7 The SFG for an eight-point DIF FFT.

The SFG for the DIF FFT is shown in Fig. 7.7. It is a left-to-right mirror image of the SFG for the DIT FFT which appeared as Fig. 7.4. Figure 7.7 also can be used for inversion, the difference being that $\{\bar{X}_n\}$ must now be entered in *natural* order, and the output will be $\{N\bar{x}_k\}$ in bit-reversed order.

Worked Example 7.5

Invert the spectrum $X(n) = 1 + j^n$ using decimation in frequency. (This was done 'directly' in Problem 5.10.)

Solution

Tabulating X_n and its conjugate, we have

n	0	1	2	3
X_n	2	$1+j$	0	$1-j$
$\bar{X}_n$	2	$1-j$	0	$1+j$
n	4	5	6	7

and $\bar{X}_n$ is to be entered in natural order. Values found at the nodes, using Fig. 7.7, are as follows:

n	$\longleftarrow$ Stage 1 $\longrightarrow$		$\longleftarrow$ Stage 2 $\longrightarrow$		$\longleftarrow$ Stage 3 $\longrightarrow$		k
	Input	Output	Input	Output	Input	*Output*	
0	2	4	4	4	4	8	0
1	$1-j$	$2-2j$	$2-2j$	4	4	0	4
2	0	0	0	4	4	0	2
3	$1+j$	$2+2j$	$2+2j$	$-4j$	-4	8	6
4	2	0	0	0	0	0	1
5	$1-j$	0	0	0	0	0	5
6	0	0	0	0	0	0	3
7	$1+j$	0	0	0	0	0	7

The stage 1 butterfly output and the stage 2 input are the same because all twiddle factors are applied to zeros. The output, $8\bar{x}_k$, is real, and the second conjugation is therefore redundant. Bit-reversal shows that the only non-zero values are $x_0 = x_6 = 1$.

SUMMARY

In Chapter 7 we have shown that a decimation process which replaces a set of N equations by sets of N, $N/2$, $N/4, \ldots, 2$ equations, all of a much simpler nature, can be used to construct a signal flow graph for the computation of a DFT. We considered the decimation-in-time FFT in some detail, and in particular showed that it gave an explanation of the origin of the factor matrices defined in Chapter 6, as their elements are identifiable from the butterfly calculations and twiddle factors used in each recursion. We have described how an alternative algebraic decimation process can be used in the frequency domain, but reverted to the concept of matrix factorization (that having by then been validated) to produce the related SFG. In Problems 7.11 and 7.12 we invite any determined reader to extend the DIT algorithm to a case in which $N > 8$, and to compute the DIF algorithm *without* recourse to the factored coefficient matrix.

PROBLEMS

7.1 (i) Show that $G(n + 4) = G(n)$, where $G(n)$ is defined by (7.5).

(ii) Show that the functions $f(n)$, $g(n)$ and $\hat{f}(n)$, defined in (7.8), (7.9) and (7.11), have period 2.

7.2 The sequence $\mathbf{x}_k, k = 0, 1, 2, \ldots, 7$ given by

$$\mathbf{x}_k = (0, \tfrac{1}{4}, \tfrac{1}{2}, \tfrac{3}{4}, 1, \tfrac{3}{4}, \tfrac{1}{2}, \tfrac{1}{4})^\mathsf{T}$$

is the sampled Λ-function which appeared in Problem 5.11. Working on the DIT FFT chart (Fig. 7.4), with initial bit-reversal, show that it has a spectrum as stated in that earlier question.

7.3 Proceeding as in Worked Example 7.2, show that

$$\mathbf{x}_k(0, 0, 1, -1 + \sqrt{2}, 2 - \sqrt{2}, -2 + \sqrt{2}, 1 - \sqrt{2}, -1)^\mathsf{T}$$

has a spectrum given by $X_1 = (-4 + 2\sqrt{2}) - 2\sqrt{2}\mathrm{j}$, $X_4 = 8 - 4\sqrt{2}$, $X_7 = \bar{X}_1$, and $X_0 = X_2 = X_3 = X_5 = X_6 = 0$.

7.4 Use the DIT FFT to invert the spectrum $\{X_n\}$ obtained in the previous example, showing that $\mathbf{x}_k$ as defined is recovered.

7.5 Invert the spectrum

$$\{X_n\} = \{0, 0, 8, 0, 0, 0, 0, 0\}^\mathsf{T}$$

using the DIT SFG, showing that $\{x_k\}$ is given by $x_k = \mathrm{e}^{\mathrm{j}k\pi/2}$, $k = 0, 1, \ldots, 7$.

7.6 From (7.19) and (7.20), obtain the expressions for $X(4p + 1)$, $X(4p + 2)$ and $X(4p + 3)$ given in (7.23), (7.24) and (7.25).

7.7 With values of $M_8^T x_k$ as given in (7.26), verify that $M_4^T(M_8^T x_k)$ is the same as the input to the third-stage butterflies on the SFG shown in Fig. 7.7, and that completion of the calculations produces the spectrum X_n in bit-reversed order. (M_4 was defined by (6.8).)

7.8 Repeat Problem 7.5, but now use the DIF SFG to invert the spectrum.

7.9 $\{x_k\}, k = 0, 1, \ldots, 7$ is given by

$$x_k = (0, 0, j, 1, 0, -j, -1, 0)^T$$

Use the decimation-in-time algorithm to show that

$$X_n = (0, (1 - j), 0, (\sqrt{2} - 1)(1 - j), -2(1 - j), (1 - j),$$
$$2(1 - j), -(\sqrt{2} + 1)(1 - j)), \quad n = 0, 1, \ldots, 7$$

Sketch the amplitude spectrum $|X_n|$. Show that Parseval's theorem is satisfied.

7.10 Use the decimation-in-frequency SFG to find the spectrum of x_k as defined in Problem 7.9 and also to invert X_n.

7.11 By following the methods used in Sections 7.2–7.5, describe how you would develop a fast DIT algorithm for a 16-point DFT.

7.12 With reference to decimation in frequency and to equations (7.18)–(7.25), if x_k is to be entered on a chart in natural order, how would you use those results to construct the SFG?

APPENDIX A

The Fourier integral theorem

In equation (1.7) of Section 1.3 we gave a description of a signal defined on an infinite range in the form of a double integral, with no explanation as to how that result was obtained. Here we offer an outline derivation, taking as starting point the Fourier *series* representation of a function defined on a *finite* range $[-l, l]$.

From (1.6) we have

$$x(t) = \frac{1}{2l} \sum_{n=-\infty}^{\infty} \left\{ \int_{-l}^{l} x(t') e^{-jn\pi t'/l} \, dt' \right\} e^{jn\pi t/l}$$

in which n is a discrete integer variable. (It is assumed that $x(t)$ satisfies the usual conditions of continuity and differentiability except, possibly, at a limited number of points.)

It is necessary that we be able to represent functions $x(t)$ defined for *all* values of t and which are not necessarily periodic. We can proceed (in a non-rigorous manner) from the expression above as follows. On putting $u = \pi/l, \Delta(nu) = u$ is the frequency spacing between consecutive harmonics in the Fourier series. Since n is integer the variable (nu) is discrete, and the previous equation can be written

$$x(t) = \frac{1}{2\pi} \sum_{nu=-\infty}^{\infty} \left\{ \int_{-\pi/u}^{\pi/u} x(t') e^{-jnut'} \, dt' \right\} e^{jnut} \Delta(nu)$$

Now consider what happens if $l \to \infty$. We may argue, first, that the interval $(-l, l)$, which implied a period $T = 2l$, becomes $(-\infty, \infty)$ and $u \to 0$, and second, that the discrete variable nu, whose values extend over a doubly infinite range, can be replaced by a continuous variable, say $nu \to w$, so that $\Delta(nu) \to dw$ and the range of w is $(-\infty, \infty)$. Using the Riemann definition of an integral as the limit of a sum, we then have

$$x(t) = \frac{1}{2\pi} \int_{-\infty}^{\infty} \left\{ \int_{-\infty}^{\infty} x(t')^{-jwt'} dt' \right\} e^{jwt} dw$$

The factor $1/2\pi$ can be removed by substituting $w = 2\pi f$, $dw = 2\pi df$. The limit process is valid if $x(t)$ *does* have a Fourier series and if $\int_{-\infty}^{\infty} |x(t')| dt'$ exists.

Either in the above form (which is equation (1.7)) or rewritten in terms of f (equation (1.8)), the Fourier integral theorem is the fundamental theorem underlying all integral transform pairs (and their discrete equivalents). The various transform pairs so validated and discussed in this text are the more significant examples of what is available.

APPENDIX B

The Hartley transform

As long as it is confirmed that a proposed integral (or discrete) transform pair leads to correct recovery of the time-domain signal on inversion, one is at liberty to design transforms with either particular applications or computational implications in mind, or both. For example, once fast algorithms were developed for processing the DFT (as described in Chapters 6 and 7) there were prompted not only considerations of improved FFT algorithms but also of modifications and alternatives to the underlying (continuous) transform pair.

A well-known example is the Hartley transform, which we here describe in both continuous and discrete forms. Related to the Fourier transform (as expected), a principal feature is that 'real' expressions are predominant. (In fast computation of the discrete form, there is some reduction in the number of multiplications required – albeit at the expense of increased additions.)

We begin by defining a continuous transform

$$H(f) = \int_{-\infty}^{\infty} x(t')[\cos(2\pi ft') + \sin(2\pi ft')]\, dt' \qquad \text{(B.1)}$$

which is the sum of a two-sided Fourier cosine transform and a two-sided Fourier sine transform. By writing $x(t')$ as the sum of its even and odd components,

$$x(t') = \tfrac{1}{2}[x(t') + x(-t')] + \tfrac{1}{2}[x(t') - x(-t')]$$

we could write (B.1) in terms of one-sided transforms and make use of the results of Sections 1.4 and 1.6 to establish that the inverse is

$$x(t) = \int_{-\infty}^{\infty} H(f)[\cos(2\pi ft) + \sin(2\pi ft)]\, df \qquad \text{(B.2)}$$

An alternative is to show directly that these two equations satisfy the Fourier integral theorem. It was shown that this can be written in the form

$$x(t) = \frac{1}{\pi} \int_0^\infty \left\{ \int_{-\infty}^\infty x(t') \cos [w(t - t')] \, dt' \right\} dw$$

which appeared previously as equation (1.9). Putting $w = 2\pi f$, $dw = 2\pi df$ and noting that $\cos [2\pi f(t - t')]$ is an even function of f, this could be re-expressed as

$$x(t) = \int_{-\infty}^\infty \left\{ \int_{-\infty}^\infty x(t') \cos [2\pi f(t - t')] \, dt' \right\} df \qquad (B.3)$$

Now suppose that $H(f)$, as defined in (B.1), is substituted into the right-hand side of (B.2), and consider the expression

$$EX = \lim_{l \to \infty} \int_{-l}^l \left\{ \int_{-l}^l x(t') [\cos (2\pi f t') + \sin (2\pi f t')] \, dt' \right\}$$
$$\cdot [\cos (2\pi f t) + \sin (2\pi f t)] \, df$$
$$= \lim_{l \to \infty} \int_{-l}^l x(t') \left\{ \int_{-l}^l \cos [2\pi f(t - t')] + \sin [2\pi f(t + t')] \, df \right\} dt'$$

on the changing the order of integration. Since $\sin [2\pi f(t + t')]$ is an odd function of f this reduces to

$$EX = \lim_{l \to \infty} \int_{-l}^l x(t') \left\{ \int_{-l}^l \cos [2\pi f(t - t')] \, df \right\} dt'$$
$$= \int_{-\infty}^\infty x(t') \left\{ \int_{-\infty}^\infty \cos [2\pi f(t - t')] \, df \right\} dt'$$

The order of integration can be changed again, and reference to (B.3) shows that $EX = x(t)$. Equations (B.1) and (B.2) therefore constitute a transform pair.

If $H(f)$ is written in terms of its even and odd components

$$H(f) = \tfrac{1}{2}[H(f) + H(-f)] + \tfrac{1}{2}[H(f) - H(-f)] = H_e(f) + H_o(f),$$

then from equation (B.1) we see that

$$H_e(f) = \int_{-\infty}^\infty x(t') \cos (2\pi f t') \, dt'$$

and

$$H_o(f) = \int_{-\infty}^{\infty} x(t') \sin(2\pi f t') \, dt' \tag{B.4}$$

The Fourier transform can be obtained from the equation

$$X(f) = H_e(f) - jH_o(f) \tag{B.5}$$

because $e^{-j\theta} = \cos\theta - j\sin\theta$. This result holds, irrespective of whether $x(t)$ is real or complex.

If $x(t)$ is real, then $H(f)$ is real and also

$$H(f) = \text{Re}\{X(f)\} - \text{Im}\{X(f)\} \tag{B.6}$$

Equation (B.6) does *not* apply if $x(t)$ is a complex signal.

The Hartley transform is more symmetrical than the Fourier transform because the transform pair, (B.1) and (B.2), have an identical structure.

The discrete form of the Hartley transform is defined by the equation

$$H(n) = \sum_{k=0}^{N-1} x_k \left[\cos\left(\frac{2\pi nk}{N}\right) + \sin\left(\frac{2\pi nk}{N}\right) \right]$$

or, with an obvious notation,

$$H(n) = \sum_{k=0}^{N-1} x_k \cos\left(\frac{2\pi nk}{N}\right) \tag{B.7}$$

and the inversion is provided by

$$x(k) = \frac{1}{N} \sum_{n=0}^{N-1} H(n) \cos\left(\frac{2\pi nk}{N}\right) \tag{B.8}$$

(In some texts, the multiplier $1/N$ appears in the definition of $H(n)$. Here it has been transferred to the inversion sum, (B.8), to facilitate comparison with the DFT as defined in (5.5).)

These equations again show complete symmetry, and no modification (such as conjugation) is needed to use any particular algorithm for both transformation and inversion. To verify that we have a transform pair we begin as we did (in the case of the DFT) in Section 5.2. In (B.7) we replace the summation integer k by m and substitute into the right-hand side of (B.8), in which k is some fixed integer, to obtain

$$EX = \frac{1}{N} \sum_{n=0}^{N-1} \left\{ \sum_{m=0}^{N-1} x_m \cos\left(\frac{2\pi nm}{N}\right) \right\} \cos\left(\frac{2\pi nk}{N}\right)$$

If the order of summation is changed,

$$EX = \frac{1}{N} \sum_{m=0}^{N-1} x_m \left\{ \sum_{n=0}^{N-1} \text{cas}\left(\frac{2\pi nm}{N}\right) \text{cas}\left(\frac{2\pi nk}{N}\right) \right\} \tag{B.9}$$

The summation over n can be written

$$\sum(m) = \sum_{n=0}^{N-1} \left\{ \cos\left[\frac{2\pi n(k-m)}{N}\right] + \sin\left[\frac{2\pi n(k+m)}{N}\right] \right\}$$

if the terms are expanded and simplified using trigonometric formulae. Considering the first term in $\sum(m)$, we can write

$$\sum_{n=0}^{N-1} \cos\left[\frac{2\pi n(k-m)}{N}\right] = \text{Re}\left\{ \sum_{n=0}^{N-1} e^{j2\pi n(k-m)/N} \right\}$$

This sum appeared in (5.7) and, by considering the sum of a geometric progression, was shown to be zero if $k \neq m$ and N if $k = m$. Similarly, the second term in $\sum(m)$ can be written

$$\sum_{n=0}^{N-1} \sin\left[\frac{2\pi n(k+m)}{N}\right] = \text{Im}\left\{ \sum_{n=0}^{N-1} e^{j2\pi n(k+m)/N} \right\}$$

in which the exponentials form another geometric progression, whose sum is

$$S_N = \frac{1 - e^{j2\pi(k+m)}}{1 - e^{j2\pi n(k+m)/N}}$$

(unless this is an indeterminate ratio). The numerator is always zero. The denominator is zero only if $(k + m) = N$, and k and m can assume values in the range $0, 1, 2, \ldots, N - 1$ only. In that case,

$$\sum_{n=0}^{N-1} e^{j2\pi n(k+m)/N} = \sum_{n=0}^{N-1} e^{j2\pi n} = \sum_{n=0}^{N-1} (-1)^n$$

which is zero as N is an even number. Hence $\sum(m)$ is zero if $m \neq k$ and has value N if $m = k$. Equation (B.9) reduces to $EX = x_k$, and that (B.7) and (B.8) are a transform pair is confirmed.

(The orthogonality relation,

$$\sum_{n=0}^{N-1} \text{cas}\left(\frac{2\pi nm}{N}\right) \text{cas}\left(\frac{2\pi nk}{N}\right) = N\delta_{mk}$$

is analogous to the orthogonality relations used when finding expressions for the coefficients $\{a_n\}$ and $\{b_n\}$ in a trigonometric Fourier series.)

APPENDIX C

Further reading

In the preface it was explained that there is an extensive literature on the subject of discrete signals, systems and transforms but that the emphasis on particular applications and the level of assumed knowledge (mathematical or otherwise) are highly variable. What follows is a short selection of titles intended to reflect this diversity.

Bateman, A. and Yates, W. (1988)   *Digital Signal Processing Design*, Pitman.

Bracewell, R.N. (1986)   *The Fourier Transform and its Applications* (2nd edn), McGraw-Hill.

Brigham, E.O. (1974)   *The Fast Fourier Transform*, Prentice Hall.

Doetsch, G. (1961)   *Guide to the Applications of Laplace Transforms*, Van Nostrand.

Kraniauskas, P. (1990)   *Transforms in Signals and Systems*, Addison Wesley.

Poularikas, A.D. and Seeley, S. (1988)   *Elements of Signals and Systems*, PWS-Kent.

Poularikas, A.D. and Seeley, S. (1991)   *Signals and Systems* (2nd edn), PWS-Kent.

Proakis, J.G. and Manolakis, D.G. (1988)   *Introduction to Digital Signal Processing*, Macmillan.

Roberts, R.A. and Mullis, C.T. (1987)   *Digital Signal Processing*, Addison Wesley.

Soliman, S. and Srinath, M.D. (1990)   *Continuous and Discrete Signals and Systems*, Prentice Hall.

Strum, R. and Kirk, D. (1988)   *First Principles of Discrete Systems and Digital Signal Processing*, Addison Wesley.

Ziemer, R.W., Tranter, W.H. and Fannin, D.R. (1983)   *Signals and Systems, Continuous and Discrete*, Macmillan.

Index